Concise Physics

H. Matyka

Edward Arnold

© H. Matyka 1987

First published in Great Britain 1987 by
Edward Arnold (Publishers) Ltd, 41 Bedford Square, London WC1B 3DQ

Edward Arnold (Australia) Pty Ltd, 80 Waverley Road, Caulfield East, Victoria 3145, Australia

Edward Arnold, 3 East Read Street, Baltimore, Maryland 21202, USA

British Library Cataloguing in Publication Data

Matyka, H.
    Concise physics.
    1. Physics—Problems, exercises, etc.
    I. Title
    530′.076   QC32
    ISBN 0-7131-3593-X

Text set in 10/11 pt Times Monophoto by Macmillan India Ltd, Bangalore 25
Printed and bound in Great Britain by The Alden Press, Oxford

Preface

Concise Physics is a systematic summary of modern A level Physics topics, which can be used to support the work done throughout a course of study and which forms a compact revision aid.

The first section deals with the important subject of study and revision techniques. It also covers the different types of questions which can be asked in an examination. The reader is encouraged to plan out a strategy so that thorough and effective use is made of the time before any paper is set.

Chapter 1 concerns itself with the basic skills required for the course, and in particular the drawing and assessment of graphs, which is an essential part of the subject.

It is introduced (as all the other chapters are) with a summary of the topics to be dealt with and a list of objectives. As the reader progresses through the text, this list may be used as an indicator of what has been covered. Particular areas of difficulty may be also marked on it for further investigation.

Points of interest and significant equations are highlighted by italicising or emboldening in the text and the use of 'boxes', which are intended to focus the attention of the reader and permit a more rapid 'take-up' of the subject matter. Worked examples and diagrams have been inserted among the topics in each chapter. The examples indicate how an answer can be presented and structured, while the diagrams emphasise the importance of illustrations for questions requiring explanations or descriptions.

Finally, I would like to thank my parents who have always given me encouragement and support in my work.

HM
1986

Contents

The technique of study and revision

Introduction

The gap which exists between GCSE and A level Physics is very wide, wider perhaps than that between A level and the first year of university. The style of questioning changes from problems which involve recall in the main to ones which require a *little thought* and *planning*. The only way to achieve success with such questions is a *thorough* understanding of the basic Physics and sufficient *practice* in answering problems of a similar nature.

To acquire a thorough understanding of fundamental Physics theory is admittedly not an easy thing to do. An important factor is something which is commonly quoted by teachers in reports, and that is *self-motivation*. The current job situation may be one source of 'motivation' but it is not usually enough, and a student must consider ways of making the subject more attractive. Side-line interests such as electronics, astronomy, various kinds of modelling and other hobbies can help. Relaxed and friendly discussion with other students and teachers can also be of use.

Organisation of private study

The planning out of a timetable for work is a very good start to a course of study. Sufficient provision needs to be made for *homework*, *note-making* and *general review*, which includes general revision and revision for specific tests.

Homework

This is the bane both of students who have to do it, and of teachers who have to mark it. Homework is best done *as soon* as it is set, as the student then has time to enquire about questions which are proving to be difficult, and thus saves time and needless frustration on the night previous to handing in—when most students do their homework.

From the start homework should be neatly presented and the questions answered in an orderly and systematic way. The various 'steps' of an answer should be reasonably brief, and written column-wise. If this is done from the beginning there is no reason why it shouldn't happen in exams.

General review

Periods of about three-quarters of an hour are usually the most profitable *without a break*. Concentration tends to fall off after this time, although longer periods of review can be withstood nearer final exams. In order to make good use of this fairly short period the student should plan what is going to be done *beforehand*. Again it is best to deal with topics in these sessions as soon as possible after the lesson(s) in which they have been dealt with, and obviously the more difficult work deserves a greater share of the general review time. It is a worthwhile idea to plan and revise the timetable for these sessions on a weekly basis.

Note making

This is perhaps the most important component of study and is determined to a certain extent by the circumstances of the student. Teachers can give very full lesson notes containing every aspect of the Physics syllabus in great detail, or they can give notes with emphasis on students making their own rough notes as the lesson progresses to be reinforced with work taken from textbooks outside the lesson. A few teachers manage to produce 'hand-outs', mainly on particularly difficult topics or on topics not covered by the students' textbooks. Notes might be provided for example on semiconductor Physics which to date has not been adequately treated in the popular Physics textbooks.

Putting the text to good use

Using the text

The 'note' form of the text presented here offers an *intermediate* stage between that of a stodgy classroom textbook and a 'snappy' revision booklet. The ideas embodied by the mathematical content of the text are important and form the 'brickwork' of A level Physics. The 'cement' which binds these ideas together is the interaction of student and teacher—who if necessary can provide reinforcement to them.

Throughout the text equations have been *boxed*. They represent, essentially, a *summary* of ideas and therefore are of importance. Only some of them need be memorised and some will have more relevance to the work in hand than others. Words have also been italicised or emboldened in the text to emphasise ideas and new terms.

The text can be used in two basic ways.

The text as a reference

Firstly, it can be used as a means for *reference* to particular topics. This is needed, for example, when doing homework. Here a student has no time to read through three or four sides of textbook wordage before the information required is pieced together, and revision booklets cannot completely supply this information. An index is included in this text which helps locate the desired item, or related topics.

The text for continual review

Secondly, it can be a *continual* aid to general review. To assist this, each chapter begins with a summary followed by a list of *objectives*—referred to as 'revision targets'. The student is thus able to mark off the objectives which have been achieved or sections which have been covered.

Particularly difficult topics may also be marked off in a similar way. *Worked examples* also feature throughout each chapter and should enable the student to tackle further problems.

Hints on making notes

The student may find it useful to make *personal notes* and *comparisons* with lesson work. The text should be studied slowly and carefully in such a way that each stage is well thought out. Ideas which have not been mentioned in lessons or which may not have been covered in enough detail can be jotted down from the text into the student's own personal notes. In this way, students will be able to concentrate on areas of the subject which are important to them.

Here are some hints on note making in general.

(1) Loose-leaf notes stored in a folder are useful as they can be added to and arranged as required throughout the course of study.
(2) Descriptions and explanations should be written out in the student's own words.
(3) Notes should be made neatly so that when they are needed, the student can actually read them.
(4) There are no limits to the use of 'boxes' or underlining and these can be used along with colour and large, clearly labelled diagrams to bring the student's notes to life. (It is a convention that Physics diagrams be drawn in black and white and thus, for the sake of exam practice, this should be adhered to even in the student's own notes.)
(5) Students' notes need to be brief but, not to the extreme that they become confusing. After all, these notes have to be read *for* revision with the aid of this text.

Revision for the exam

It may seem strange to discuss revision at the beginning of a series of notes on A level Physics. The reason is that revision should start *as soon* as the course does. There is no way a student can make a thorough revision of a two-year course in the last week before the exam. This is the purpose of the general review which acts as a method of revision on a weekly basis and reinforces work already done for material to be covered in the future.

For the final two months of study, however, it is wise to plan out a *timetable for revision*, making sure that a variety of topics is to be covered so that concentration and interest are both maintained.

A student can often benefit from revising in a small group with others. It is also important that he be aware of what exactly is expected of him in the exam. That is the type of paper, the length of time allowed, the sort of questions which tend to be asked, etc.

The student should make a determined effort to try out exam papers *under exam conditions* at home and to go through as many questions from past papers as possible—even if some are only outline answers.

The examination

Few students recognise the importance of *examination technique*. It is something which cannot guarantee a pass but which will offer the candidate a slight *marking edge*.

Firstly, it is important to know or at least to make sure of the instructions appropriate to a particular paper. For example in a section requiring answers to only two of the questions, there will always be one candidate who answers all four.

Secondly, it is important to spend a few minutes *reading* through the questions, and perhaps making a number of brief notes about suitability for answering and maybe some memorised formulae appropriate to the questions which might evade recall later on. (Multiple choice papers should not be approached in this way.)

Thirdly, the student, who should already be aware of the amount of time he can devote per question, can now start to answer those questions he has decided upon, taking the easiest ones first. Most A level papers now have the marks allocated for different parts of an answer alongside the question. These indicate the amount of time which needs to be spent on that part and the amount of detail required. For example the examiner will expect more of an answer for a part allocated 75 % of the question marks than from a part allocated only 10 %.

Not overspending time on the easy questions also leaves enough time for consideration of the more difficult questions.

The *fourth point* is of great importance: the student should *read the question he is answering carefully*. This may seem a logical thing to do in the first place, but many, many students lose marks through not having read the question properly.

For example a question might ask for the statement of a particular physical law. There will always be students who write down an equation and leave it at that. Now if the question had asked for the equation to a particular law the answer would be right, but as it is their answer is wrong.

Another example is the description of an experiment as answer to a 'define the . . .' question. This can lead to an incorrect answer and hence lost marks and also a *wastage of time*.

Types of exam question

There are four basic types of exam question used in Physics papers. These are:

(1)   multiple choice,
(2)   data analysis,
(3)   comprehension,
(4)   general theory.

They require slightly different approaches.

Multiple choice

Here the student is often rather restricted by time per question. There is, however, the benefit of a statistical chance of selecting the right answer without actually knowing it.

The questions are best treated 'as they come'. With the time restriction in mind, the student must work through the paper ignoring the more difficult questions and doing them on a second, third or fourth 'run'. Needless to say an attempt should be made at *all* the questions by the end of the paper.

Individual questions also have variations. Generally a question will have five possible answers and usually two or three are ridiculous if not very dubious. The choice then must fall among the remainder. Other questions involve only three answers with the complication that a combination of right answers is possible. Although this may seem tricky at first (and these questions are unpopular), by the elimination of the very dubious answer a number of the combinations can also be dispensed with.

Data analysis

This basically involves the rearrangement of an equation or equations to obtain a suitable form for a graphical treatment of the data supplied in the question. Neatness of presentation, orderliness and the observation of the rules for drawing graphs are required here.

Comprehension

In this case, a passage of scientific interest is quoted, containing material which may or may not be novel to the student. Having read this passage the student has to answer a few questions on it, applying the basic understanding of Physics which should have been acquired. As a check, the section corresponding to the question should be re-read. In *sketching* graphs the usual rules do not apply. Instead the graph is sketched onto lined not graph paper and although the axes are labelled with quantity and units, they are not numbered. (The rough values at a particular feature may, however, be shown.) Theory outside that of the passage which cannot be considered as basic, or any assumed data should *not* be involved in the student's answers. These must also be expressed in the student's own words and not copied from the text. It may be useful in particular cases to explain an answer with the aid of a diagram.

A major problem that students have is the inability to express themselves either orally or in writing. Thus, a comprehension exercise can be difficult to cope with, and some practice is essential.

General theory

These can either be short, one part questions or lengthy ones with many parts. Some hints on handling these have been given in the section headed *The examination*. The answers must be neat and presented in an orderly fashion. Calculations should involve the following steps:

(1) name of the theory and perhaps a statement of the law being applied,
(2) rearrangement of the basic equation until the required term becomes the subject,
(3) substitution of the given values and the answer, not forgetting any units.

Clearly labelled, large, neat diagrams are always an aid to explanation and show that the student has confidence in what is being written.

Final point

As in any course of study, the student must take note of, and act upon, the comments and advice of teachers and instructors who can help him, *along with notes*, to cover any points of weakness.

1
Basics

Introduction

This chapter concerns itself with the basics of presenting physical quantities and relationships.

It begins with a consideration of the system of units and the notation which are used to represent the size or magnitude of a quantity. The dimensions of quantities can then be used to check equations. This is followed by a section on graphs which is introduced with the procedure for drawing them and a discussion of the linear graph. The gradient is one of the measurements which can be made from this important type of graph which can also be applied to the verification of equations. Different types of graph are looked at next, as well as the way in which they can be transformed to a linear form.

The chapter ends with an investigation of the accuracy of measurements and the practical procedures for their determination.

Revision targets

The student should be able to:

(1) Apply the correct units and unit prefixes to physical quantities.
(2) Transform decimal notation to standard notation.
(3) Change the subject of an equation and to check its consistency as regards dimensions.
(4) Distinguish between expressions containing only units and those containing only physical quantities.
(5) Transfer information between graphs, algebraic expressions and statements.
(6) Determine the gradient of a graph in the correct units, and recognise its physical significance.
(7) Recognise the major types of graph.
(8) Verify a predicted relationship for a set of data by drawing a linear graph and make other deductions from it.
(9) Appreciate the meaning of proportionality.
(10) Recognise possible sources of error in experimental work.
(11) Apply discretion as regards the accuracy to which the answer to a calculation is given.
(12) Describe briefly but accurately the procedure followed in an experiment for class and examination purposes.
(13) Appreciate the need for caution when taking readings to ensure that the measurements are as reliable as possible.

Physics

Physics is the study of the properties of **matter** and **energy**. This study takes the form of observation, measurement and consideration of how things work and why. Some useful terms in Physics include:

(1)   a *fact*—an experimental observation which can be repeated,
(2)   a *theory*—a set of ideas based on experiment,
(3)   a *law*—sums up ideas of great importance simply and clearly (it can usually be put in the form of an equation).

The units of the **Système International** (SI) are now used to give the value of a property.

Units

Basic SI units

Other units

Property	Name of unit	Symbol
Length	Metre	m
Mass	Kilogramme	kg
Time	Second	s
Electric current	Ampère	A
Thermodynamic temperature	Kelvin	K
Amount of substance	Mole	mol
Luminous intensity	Candela	cd
Angle	Radian	rad
Solid angle	Steradian	sr

These units are based on certain standards. For example the kg is the mass of a standard platinum cylinder kept in Paris.

Standard notation

This is an easy way of writing down very large or very small numbers, by using powers of ten. (Every number can be expressed in terms of the number ten.) For example:

$$20 = 2 \times 10$$

$$3000 = 3 \times 10 \times 10 \times 10$$

$$0.35 = \frac{35}{10 \times 10}$$

Number	Standard notation
$30 = 3 \times 10$	3.0×10
$200 = 2 \times 10 \times 10$	2.0×10^2
$7000 = 7 \times 10 \times 10 \times 10$	7.0×10^3
$89.3 = 8.93 \times 10$	8.93×10
$5850 = 5.85 \times 10 \times 10 \times 10$	5.85×10^3
$0.3 = 3/10$	3.0×10^{-1}
$0.000\,024 = \dfrac{2.4}{10 \times 10 \times 10 \times 10 \times 10}$	2.4×10^{-5}

When multiplying numbers with powers of ten in them, the powers *add*. When dividing numbers with powers of ten in them, the powers *subtract*.

A good example of the useful application of standard notation is Avogadro's constant. In one mole of an element there are 6×10^{23} atoms. In long hand this would be six followed by twenty-three zeros.

Calculations with standard notation

Some examples:

$$2 \times 10^4 \times 3.5 \times 10^{-3} = 2 \times 3.5 \times 10^{4+(-3)} = 7 \times 10 = \underline{70}$$

$$\frac{5 \times 10^7}{2.5 \times 10^2} = \frac{5}{2.5} \times 10^{7-(-2)} = \underline{2.0 \times 10^9}$$

$$\frac{200 \times 600\,000}{0.002} = \frac{2 \times 10^2 \times 6 \times 10^5}{2 \times 10^{-3}} = \frac{12 \times 10^7}{2 \times 10^{-3}} = \underline{6.0 \times 10^{10}}$$

Prefixes for SI units

Name	Symbol	Value
Mega	M	10^6
Kilo	k	10^3
Centi	c	10^{-2}
Milli	m	10^{-3}
Micro	μ	10^{-6}
Nano	n	10^{-9}
Pico	p	10^{-12}

For example:

$$1\,\text{mm} = 1 \times 10^{-3}\,\text{m} \qquad 1\,\mu\text{m} = 1 \times 10^{-6}\,\text{m}$$

The basic SI units are used to define *others*. For instance the unit of force, called the **newton** whose symbol is N, can be written as

$$1\,\text{N} = 1\,\text{kg}\,\text{m}\,\text{s}^{-2}$$

(Note: It is usual to put s^{-2} rather than $1/s^2$ or even $1/s/s$.)

Usually in Physics, when multiplying units or properties, the symbol for multiplication is not written. Sometimes it is replaced by a dot.

Method of dimensions

This is a method of checking formulae and equations by making sure that the basic SI units either side of the 'equals' sign are the same. The **dimensions** of a quantity are defined as the powers to which the basic SI units have to be raised, in order to express the units of that quantity. Dimensions are normally written in square brackets.

For example, in an exam, a student writes the equation for force as

force = mass × distance × time

Unit	Dimension
m	[L]
kg	[M]
s	[T]
A	[A]
K	[θ]

when the right hand side of this expression is written in the form of dimensions, it gives

$$[M]\,[L]\,[T]$$

However, the dimensions of the newton are $[M]\,[L]\,[T]^{-2}$, so the equation the student has written is *wrong*. The method cannot be applied to dimensionless quantities. Thus, if it is used to *suggest* a relationship, a proportionality sign is required. Units and properties should not be mixed up in the same formula but must be kept in *separate* equations.

Graphs

Graphs are very helpful in Physics, because they show how results from an experiment are *related*. They are drawn according to the following procedure.

(1)   Axes are drawn in ink. They are labelled with the quantity being plotted and its **units**. The axes are arrowed to show that the quantity is increasing in value.
(2)   The **scale** of the graph should make it easy to read and use. For example 5, 10, 20 'somethings' per cm are good scales, whereas 3, 7, 8, 9 'somethings' per cm are bad scales.
(3)   Axes are numbered in pencil first, then *neatly* in ink.
(4)   The graph should be as *large* as possible.
(5)   Points on the graph are drawn in pencil, as *crosses*, and a smooth pencil line or curve is then positioned through as many and/or as near to as many crosses. The crosses are never joined up one by one like a 'dog's hind leg'.
(6)   Every graph should have a title.

Linear graphs

A straight line or **linear graph** is the most useful type of graph in Physics. It clearly indicates that there is a definite relationship between one changing quantity plotted along one axis and another changing quantity plotted along the other axis.

These quantities are called **variables**, because they are changing. Quantities which do not change are called **constants**. If two variables, variable (1) and variable (2), which are related by the following equation are plotted against each other, a straight line graph as shown in Fig. 1.1 will be produced.

variable (1) = [constant (A) × variable (2)] + constant (B)

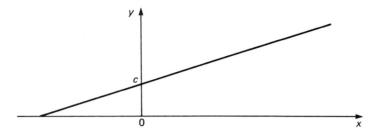

Figure 1.1

In mathematics the above expression is often written as

$$y = mx + c$$

where x and y are the variables and m and c are the constants.

m is called the **gradient** or the slope of the graph. The value of the gradient indicates the rate of change of y with respect to x. That is, if m has a high value, then y changes very markedly for only a small change in x.

The constant c is the point at which the graph cuts the y-axis. It gives the value of y when x is zero.

In mathematics, both m and c are just numbers. However, in Physics these quantities have physical meaning and usually have *units*. The units of m are given by the units of y divided by those of x. The units of c are the units of y.

Finding the gradient

In order to determine the gradient of a linear graph a 'gradient triangle' is drawn to the graph, and its sides are labelled with the change in y and the change in x, respectively, that those sides are equivalent to. To minimise error the gradient triangle should be made as large as possible. See Fig. 1.2.

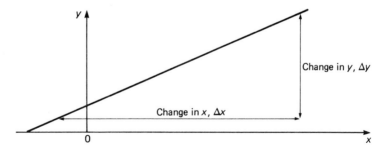

Figure 1.2

$$\text{gradient } m = \frac{\text{change in } y}{\text{change in } x} = \frac{\Delta y}{\Delta x}$$

When the changes in y and x are made infinitesimally small, m is written as

$$m = \frac{\mathrm{d}y}{\mathrm{d}x}$$

This is called the **differential** of y with respect to x. It is *not* a fraction, since an infinitesimally small value divided by another infinitesimally small value would always give one.

The gradient along a linear graph is the same at each point on the graph.

Gradients for non-linear graphs

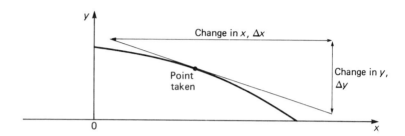

Figure 1.3

The gradient along a non-linear graph or curve changes from point to point. However, the gradient at one point is still determined by drawing and labelling a large gradient triangle as for the linear graph. The hypotenuse of the triangle is drawn so that it is a tangent at the point on the curve under investigation. That is, it 'follows' the curve as well as possible around that point. See Fig. 1.3.

Negative gradient

If the value of y on a graph is decreasing as the value of x increases, the graph has a **negative gradient**. Different parts of a curve may well have positive or negative gradients.

Data analysis

The results from an experiment are most easily investigated when they are presented in a graphical form. Trends and characteristic features in a 'string' of data are more apparent on a graph and it is possible to make numerical deductions from them.

Verification of formulae

Sometimes the relationship between the changing quantities in an experiment can be *predicted*. A method for verifying the relationship or law is to convert it to a *linear* form:

variable (1) = [constant (A) × variable (2)] + constant (B)

The values of the variables (1) and (2) then need to be calculated *from* the results of the experiment. If a graph is then drawn of variable (1) versus variable (2), a straight line *should* be produced. When the predicted relationship is *wrong*, the linear form will be wrong for the data and *no straight line* will be obtained.

Other types of graph

See Fig. 1.4.

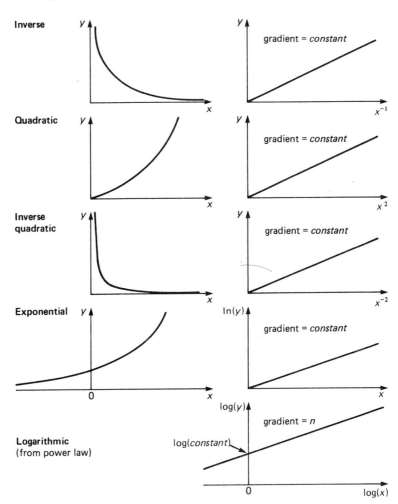

Figure 1.4

Inverse

General formula:

$$\boxed{y = constant/x}$$

Or: $y = constant \cdot x^{-1}$

For a linear graph,

variable (1) $= y$ and   variable (2) $= x^{-1}$

constant $(A) = constant$ and   constant $(B) = 0$

(line passes through origin)

Quadratic

General formula:

$$\boxed{y = constant \cdot x^2}$$

For a linear graph,

variable (1) $= y$ and   variable (2) $= x^2$

constant $(A) = constant$ and   constant $(B) = 0$

(line passes through origin)

Inverse quadratic

General formula:

$$\boxed{y = constant/x^2}$$

Or: $y = constant \cdot x^{-2}$

For a linear graph,

variable (1) $= y$ and   variable (2) $= x^{-2}$

constant $(A) = constant$ and   constant $(B) = 0$

(line passes through origin)

Exponential

General formula:

$$\boxed{y = e^{constant \cdot x}}$$

Taking *natural* logs,

$$\ln(y) = constant \cdot x$$

For a linear graph,

variable (1) $= \ln(y)$ and   variable (2) $= x$

constant $(A) = constant$ and   constant $(B) = 0$

(line passes through origin)

Note: ln denotes $\log_e$.

Power law (application of logarithms)

General formula:

$$\boxed{y = constant \cdot x^n}$$

Taking logs,

$$\log(y) = \log(constant \cdot x^n)$$

Thus, $\log(y) = n\log(x) + \log(constant)$

For a linear graph,

variable $(1) = \log(y)$ and variable $(2) = \log(x)$

constant $(A) = n$ and constant $(B) = \log(constant)$

Note: log denotes $\log_{10}$.

Example 1.1

A thermocouple, which is a particular type of thermometer (discussed later in the text), produces the following series of emf values E, for various temperature differences θ across its junctions.

$\theta(°C)$	emf (mV)
50	0.43
100	0.79
150	1.10
200	1.36
250	1.54
300	1.69
350	1.77
400	1.80
450	1.78
500	1.70
550	1.54

The relationship between E and θ is predicted to be of the form

$$E = a\theta + b\theta^2 \qquad (a \text{ and } b \text{ are constants})$$

(i) Draw a graph of E versus θ and determine the temperature at which the maximum emf occurs. (This temperature is called the **neutral temperature**.)

(ii) Draw a suitable graph enabling the verification of the above relationship for the given data and use it to determine the values of the constants a and b.

(i)

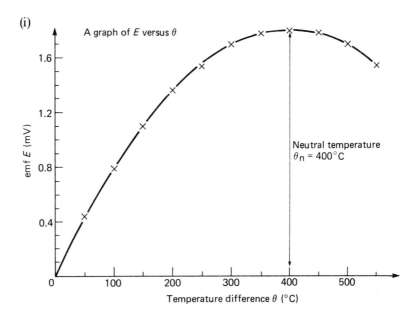

Figure 1.5

The neutral temperature as determined from the graph in Fig. 1.5 is 400°C.

(ii)   In order to verify the relationship for the given data and determine the constants a and b, a straight line graph of the following form can be drawn:

$$\text{variable (1)} = [\text{constant } (A) \times \text{variable (2)}] + \text{constant } (B)$$

The expression for E has to be rearranged in order to get it into a linear form:

$$E = a\theta + b\theta^2$$

Dividing through by θ gives

$$\frac{E}{\theta} = a + b\theta$$

Finally,

$$\frac{E}{\theta} = b\theta + a$$

Thus,   variable (1) $= E/\theta$ and   variable (2) $= \theta$

constant (A) $= b$ and   constant $(B) = a$

Now, if E/θ is plotted against θ, a straight line graph should result whose gradient is equal to b and whose intercept on the y-axis is a.

θ (°C)	E/θ (μV °C^{-1})
50	8.6
100	7.9
150	7.3
200	6.8
250	6.2
300	5.6
350	5.1
400	4.5
450	4.0
500	3.4
550	2.8

Note the prefix change and the units for E/θ.

All the data lie on or very close to the straight line and thus the predicted relationship is verified for the given data.
   The gradient of the graph in Fig. 1.6 is

$$\frac{\Delta y}{\Delta x} = b = -\frac{4.5 \times 10^{-6}}{400} = \underline{-1.13 \times 10^{-8} \text{ V}°\text{C}^{-2}}$$

The intercept on the y-axis is

$$a = \underline{9 \times 10^{-6} \text{ V}°\text{C}^{-1}}$$

(Note the units for a and b, and the use of standard notation).

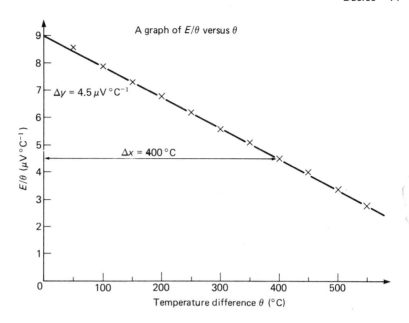

Figure 1.6

Exponential decay

When two variables y and x are related by the expression

$$y = a \cdot e^{-b \cdot x}$$

where a and b are constants, then y is said to **decay** exponentially with x. The graph of y versus x is shown in Fig. 1.7.

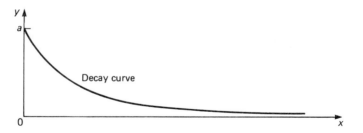

Figure 1.7

Processes which exhibit exponential decay include the *decay* of radioactive substances and the *discharge* of capacitors.

Proportionality

When two variables x and y are related by the equation

$$y = constant \cdot x$$

then y is said to be **directly proportional** to x. When they are related by the equation

$$y = constant/x$$

then y is said to be **inversely proportional** to x. The two types of proportionality are represented as follows:

$$y \propto x \quad \text{and} \quad y \propto 1/x$$

For example, if y is inversely proportional to x^4, then

$$y \propto 1/x^4$$

Errors

There are two main types of error: **random** and **systematic**.

Random error arises when repeated measurements of one quantity give different values; for example the slightly different times measured for the descent of a weight when it is repeatedly dropped through a distance of 2 m.

Systematic error arises when all the measurements of one quantity are affected *equally* by some influence. For example, if the times taken for a weight to fall through different distances are measured with a timer that does not go back to zero but to the 2 ms mark, then a systematic error occurs.

Dealing with errors

Random errors can be dealt with by repeating the reading many times and obtaining the **average** or **mean** value of the measurement. This improves the reliability of the reading.

Systematic errors are best treated by comparing readings with a **standard** instrument or by *checking* them against the results that theory indicates should be produced. They are hard to uncover and it is a case of analysing an experimental method carefully and being aware of the limitations of the apparatus.

Decimal places and significant figures

The first non-zero digit in a number is called the first **significant figure**. For example the following numbers all have three significant figures.

596   68.7   0.456   0.000 034 2

When measurements determined in experiments are used in a calculation, the accuracy to which the answer is given should *match* the accuracy of the measurements. This rule should be applied to all types of *calculation*. For example, if the time taken for a trolley to travel with uniform motion a distance of 0.5 m is 3 s, its speed can be calculated as follows:

$$v = \frac{d}{t} = \frac{0.5}{3} = 0.166\,666\,666\,6 \, \text{m s}^{-1}$$

The answer has been written to ten decimal places and it assumes that the measurements are accurate also to ten decimal places, which they are *not*. In this case the answer should be written correct to, at the *most*, two decimal places, that is:

$$v = \underline{0.17\,\text{m s}^{-1}}$$

Similar care is required with calculations involving large numbers. For example, if the speed of light is given as $3 \times 10^8 \, \mathrm{m\,s^{-1}}$ and the time it takes to cover a distance of $1 \times 10^{14} \, \mathrm{m}$ is t, then t can be calculated as follows:

$$v = \frac{d}{t}$$

then

$$t = \frac{d}{v} = \frac{1 \times 10^{14}}{3 \times 10^8} = 333\,333.333 \, \mathrm{s}$$

The answer should be given at the *most* to two significant figures, that is:

$$t = 330\,000 \, \mathrm{s} = \underline{3.3 \times 10^5 \, \mathrm{s}}$$

Practical Physics

During a Physics course, experiments are used to emphasise points of *theory* and to enable students to develop basic *observational* skills. Experiments need to be written up neatly and clearly, as follows.

(1) **Title** This should be as simple as possible and indicate clearly what the *purpose* of the experiment is.
(2) **Apparatus list** This should indicate the main apparatus used and particularly any which is rarely used.
(3) **Diagram** The diagram should be drawn in pencil, clear, large and neatly *labelled*. A ruler and compass must be applied where necessary rather than free hand. If the diagram shows how the apparatus is initially set up it can save a lot of descriptive work in the method.
(4) **Method** This is best written as a series of numbered steps. These steps should be briefly written up, but should accurately describe what *has* been done. Personal names and adjectives must not be included in the write-up (i.e. John, Mary, he, we, they, etc.).
(5) **Results** Where possible these ought to be presented in a table and on a graph.
(6) **Conclusion** This should contain a *summary* of what has been deduced from the experiment. It may also discuss some of the theory behind the experiment and possible sources of error.

Practical Physics in the exam

Questions requiring a description of an experiment in the written exam should be answered by applying the usual rules for the writing up of experiments, except that:

(1) The method needs to be written up in the present tense, i.e. what *is* done rather than what has been done.
(2) No *numerical* results should be given. It is a common mistake for students to write 'imaginary' or 'made-up' results. The trends expected in the results for the experiment can, however, be shown by *sketching* a graph.

A good, clear diagram is essential in such descriptions since it can save time in writing up the method. The numbered steps in the method allow the examiner to see more clearly the actions performed in the experiment.

Performing an experiment in the exam

A practical examination generally consists of the taking of a set of readings using a particular set of apparatus and the graphical treatment of the results.

Careful consideration of how best the apparatus can be set up needs to be made, and the student should divide up the available time so that there is sufficient for the treatment of the results. The write-up of the experiment should follow the usual rules for the writing up of practicals, except that:

(1)   The detail to which the method is written up should correspond to the requirements of the syllabus. It should *not* be a word-for-word copy of the instructions printed on the exam paper.

(2)   The method should also contain the steps taken to ensure that the results are as *reliable* as possible. The student should consider what is actually being measured. For example, if a distance is being measured, how can the distance be made more accurate? If it is a large distance then a metre rule needs to be applied rather than the successive use of an ordinary ruler. Perhaps a diameter is required; then calipers might be of use. The precautions taken are determined by what the required measurement is.

(3)   All readings made need to be recorded. The student should investigate the whole range of possible measurements, taking the readings at *regular* intervals. Repeated readings must be made if possible to provide *averages*.

(4)   The apparatus should not be taken apart before the write-up and treatment of the results. This means that, for example, the student can go back and take further readings—perhaps to investigate a particular area on a graph which requires investigation. (Readings at the extremes of ranges tend to be less accurate than those at the middle.)

(5)   It is quite easy to get flustered in the practical exam and spoil the results. To avoid this the student should have had *practice* practical exams and should know how the paper is to be tackled. The supervisors during the exam may be asked for guidance in cases of severe difficulty, but whether this can be given is determined by their discretion and the examination requirements.

Example 1.2

Describe an experiment to investigate the relationship between the length of a simple pendulum and the time taken for it to perform one complete oscillation.

An experiment to investigate the oscillation of a simple pendulum

Apparatus   1 m length of string with small lead mass attached to it, stop watch, metre rule, stand and clamp.

Method
(1)   The apparatus is set up as shown in the diagram.
(2)   The pendulum is held between 2 small blocks of wood which are clamped together, enabling accurate determination of the length of the pendulum.
(3)   A chalk mark is made at the side of the bench to enable the position of the pendulum to be judged more easily.

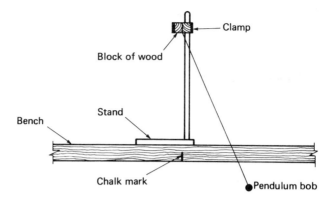

Figure 1.8

(4)   The length of the pendulum is altered in regular steps of 0.1 m and for each length the time taken for 50 complete swings is measured three times using the stop watch. (The displacement of the pendulum is kept small.) The values for length and time are recorded.

Results
The *average* time taken for *one* complete swing, T, is calculated and tabulated along with all the readings from the experiment.
   A graph of T^2 versus L, where L is the length of the pendulum, is then plotted. [See simple pendulum theory.] This should be a straight line graph.

Conclusion
The straight line graph shows that the time taken for one complete oscillation of the pendulum, squared, is directly proportional to the pendulum length, since the line passes through the origin.

   Example 1.2 concerns itself with the written response to an exam question, but it also shows how a student might approach the method for a practical exam. The two measurements required in this experiment are length and time, and steps to improve the reliability of these are taken in the method. This sort of consideration plays an important role in many practical exam situations.

2
Mechanics

Introduction

This branch of Physics deals mainly with forces and the effects of their application to bodies. The chapter starts with motion and the different types of force. The effect of a force on a body's state of rest and motion are then successively dealt with. This is followed by an investigation of circular motion and gravitation. Finally the application of a force to cause oscillatory motion and external influences upon it are treated.

Revision targets

The student should be able to:

(1) Interpret graphs showing the motion of a body.
(2) Apply the equations of uniformly accelerated motion to simple problems.
(3) Add, subtract and split up vectors, particularly velocity and force.
(4) Appreciate the difference between mass and weight.
(5) Investigate the equilibrium of a body by determining the moments acting upon it.
(6) State Newton's laws of motion and apply the second law to numerical problems.
(7) Define linear momentum and state the principle of conservation of linear momentum.
(8) Experimentally investigate Newton's second law of motion and the conservation of linear momentum.
(9) Describe the general laws and nature of friction.
(10) Define and give equations for work, energy and power and apply them to numerical problems.
(11) State the law of conservation of energy.
(12) Define and interrelate angular velocity, frequency, and period of rotation.
(13) Determine the acceleration of a body moving in a circle.
(14) Define centripetal force.
(15) State Kepler's laws and Newton's law of gravitation.
(16) Apply Newton's law of gravitation to various problems involving orbiting bodies.
(17) Describe how and why the acceleration due to gravity varies.
(18) Define terminal speed and gravitational potential.
(19) Define moment of inertia and appreciate the factors which determine its value.
(20) Define angular momentum and state the conservation law for it.

(21) Define simple harmonic motion and apply it to cases, particularly those involving the simple pendulum and loaded spring.
(22) Define damping and resonance and give examples of the latter.
(23) State Hooke's law and describe the effect of applying a load in order to stretch different types of material.
(24) Distinguish between stress and strain and define Young's modulus of elasticity.
(25) Describe an experimental determination of Young's modulus for a wire.
(26) Discuss the work done and energy stored for a stretched wire and apply the theory to solve numerical problems.

Scalar and vector quantities

There are two basic types of quantity in Physics.

A **scalar** quantity is one that has only size (magnitude), for example time or mass.

A **vector** quantity is one that has both size and direction. For example a boy walks 15 m eastwards. A distance moved in a particular direction is called a *displacement*. Since it has both a size and a direction, displacement is a vector.

Rectilinear motion

This type of motion is the movement of a body in a straight line. In order to investigate the motion fully some basic definitions are required.

Speed

This is the rate of change of distance with time. A value of the average speed is determined by applying the equation

$$\text{average speed} = \frac{\text{distance travelled}}{\text{time taken}}$$

It has units of $m\,s^{-1}$ and is a scalar quantity because it does not have a direction.

Velocity

This is the rate of change of distance with time in the direction of motion. Since it has direction it is a vector. Its magnitude is the speed of the body and it has units of $m\,s^{-1}$.

Acceleration

This is the rate of change of velocity with time. Since velocity is a vector so is acceleration. Its magnitude is obtained from the equation

$$\text{acceleration} = \frac{\text{change in velocity}}{\text{change in time}}$$

Acceleration has units of $m\,s^{-2}$.

Summary

Quantity	Type of quantity	Units
Distance moved	Scalar	m
Displacement	Vector	m
Speed	Scalar	$m/s = m s^{-1}$
Velocity	Vector	$m/s = m s^{-1}$
Acceleration	Vector	$m/s/s = m s^{-2}$

Graphical representation

The motion of a body travelling in a straight line can be expressed graphically in the form of a distance–time graph or a velocity–time graph. The *gradient* of the former is given by

$$\frac{\text{change in distance}}{\text{change in time}}$$

which is the magnitude of the velocity of the body. The *gradient* of the latter is given by

$$\frac{\text{change in velocity}}{\text{change in time}}$$

which is the acceleration of the body.

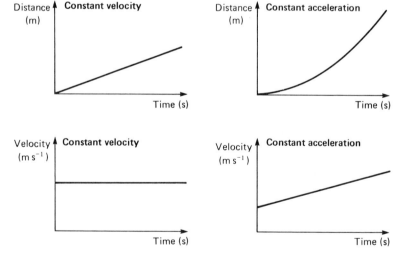

Figure 2.1

The gradients of the graphs in Fig. 2.1 can be worked out by applying the equation

$$\text{gradient} = \frac{\text{change in } y}{\text{change in } x}$$

If the changes in y and x are made infinitesimally small, then the gradients can be expressed in the form of *differentials*.

$$\text{velocity } v = \frac{ds}{dt} \quad \text{and} \quad \text{acceleration } a = \frac{dv}{dt}$$

Here s has been used as the symbol for distance travelled. The student should note that because the gradients are actual physical quantities, *they have units.* (In fact most gradients and intercepts of graphs in Physics have units, which is in sharp contrast to the graph work done in mathematics.)

Areas under graphs

In practice the area under a graph may be found by counting the number of squares under the graph. Once again the area will probably have units given by the product of the units of the quantities along the axes. A similar process is to divide the area into vertical strips and add up the areas of these strips. For example for a velocity–time graph the strips might be Δt wide and v tall. Such an adding up process using infinitesimally thin strips is called, in mathematics, **integration**. It is written as follows, with the 'stretched out' S simply standing for sum:

$$\text{area under velocity–time graph} = \int v \, dt = \text{distance travelled}$$

Similarly, it can be shown that the area under an acceleration–time graph gives the average velocity of a body.

Equations of uniformly accelerated motion

In uniformly accelerated motion the acceleration of the body is a *constant*, i.e. it stays the same all the time.

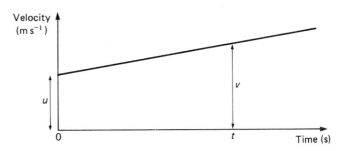

Figure 2.2

Let the initial velocity of a moving body be u and the velocity at a later time t be v. Then from the graph in Fig. 2.2,

$$v = u + \frac{\text{(change in velocity)}}{\text{change in time}} \text{(change in time)}$$

So

$$\boxed{v = u + at}$$

The distance travelled s is given by the *area* under the velocity–time graph. Thus,

$$s = ut + \frac{(v - u)t}{2}$$

or
$$s = \frac{(u + v)t}{2}$$

Since $v = u + at$ then
$$s = \frac{(u + u + at)t}{2}$$

So
$$s = ut + \tfrac{1}{2}at^2$$

Finally, from $v = u + at$
$$t = \frac{v - u}{a}$$

and substituting for time in the expression for distance:
$$s = \frac{(u + v)(u - v)}{2a}$$

and therefore
$$2as = v^2 - u^2$$

An example of uniformly accelerated motion is that of a body falling under the action of gravity towards the Earth's centre. Neglecting air resistance which is a force acting upwards on the body, the mean value of the **acceleration due to gravity** (also called the **acceleration of free fall**) is

$$g = \underline{9.81 \text{ m s}^{-2}} \quad \text{or approx.} \quad g = 10 \text{ m s}^{-2}$$

The value of g varies geographically and thus a mean value is used. All bodies fall at the same rate when dropped from the same height in a vacuum. This was demonstrated in one of the first experiments performed on the moon. A feather and a weight, dropped from the same level, both hit the moon's surface at the same time.

Example 2.1

Figure 2.3 shows a velocity–time graph for a body moving in a straight line. Determine the initial acceleration, final acceleration and the total distance covered by the body.

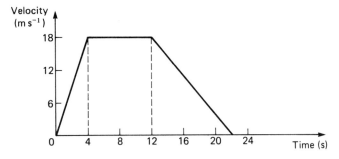

Figure 2.3

Acceleration is determined from the *gradient* of this graph. Thus,

$$\text{initial acceleration} = \frac{\Delta y}{\Delta x}$$

$$= \frac{18}{4} = \underline{4.5 \text{ m s}^{-2}}$$

$$\text{final acceleration} = \frac{\Delta y}{\Delta x}$$

$$= \frac{-18}{10} = \underline{-1.8 \text{ m s}^{-2}}$$

i.e.   deceleration $= 1.8 \text{ m s}^{-2}$

Total distance covered is given by the *area* under the velocity–time graph.

$$\text{distance covered} = \tfrac{1}{2}(4 \times 18) + (8 \times 18) + \tfrac{1}{2}(10 \times 18)$$
$$= \underline{270 \text{ m}}$$

Example 2.2

A car travelling at a speed of 29 m s^{-1} brakes to a stop in 11 s. What is its acceleration? How far does it travel during this period?

Applying $v = u + at$, then

$$a = \frac{v - u}{t} = \frac{0 - 29}{11} = \underline{-2.6 \text{ m s}^{-2}}$$

How far does it travel during this period?

Applying $s = \frac{(v + u)t}{2}$, then

$$s = \frac{(0 + 29)11}{2} = \underline{160 \text{ m}}$$

Note that a negative acceleration is a positive deceleration.

Example 2.3

An aircraft flies by an anti-aircraft missile base at a speed of 120 m s^{-1} in level flight. If a missile inclined at 45° to the ground is fired at that same instant with an acceleration of 75 m s^{-2}, how long will it take to reach the aircraft? Assume that the missile travels along a straight trajectory in the direction of travel of the aircraft.

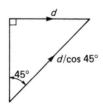

Figure 2.4

In Fig. 2.4, distance travelled to intercept:

by aircraft $= d$
by missile $= d/\cos 45°$

If $v' =$ speed of aircraft, then

$$v' = d/t$$

Applying $s = ut + \tfrac{1}{2}at^2$ for the missile, then

$$\frac{d}{\cos 45°} = \tfrac{1}{2}at^2$$

and
$$\frac{v't}{\cos 45°} = \tfrac{1}{2}at^2$$

Therefore
$$t = \frac{2v'}{a\cos 45°} = \frac{240}{53}$$

Thus,
$$t = \underline{4.5\ s}$$

Force

A **force** is any action which alters or tries to alter the state of rest of a body or its uniform speed in a straight line. It has a direction and therefore is a *vector* quantity. The **newton** (symbol **N**) is the unit of force. It is the force required to accelerate a mass of 1 kg through one metre per second per second.

Addition of vector quantities

In diagrams, a vector is represented by an arrow. The arrowhead indicates the direction of the vector and the length of the arrow represents the magnitude of the vector. For example, if a scale of 1 cm to every 1 m is used, a displacement of 10 m eastwards is drawn as an arrow of length 10 cm as shown in Fig. 2.5.

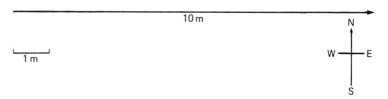

Figure 2.5

When vectors are added, their sum is called the **resultant** vector. *Parallel* vectors acting at a point add or subtract like ordinary numbers.

Example 2.4

Two students together push a large box in the same direction. If each applies a force of 50 N, what is the resultant force on the box?

Figure 2.6

See Fig. 2.6. The resultant force is
$$F = 50 + 50 = \underline{100\ N}$$

Example 2.5

A car travels along a motorway at 25 m s^{-1} eastwards. How would the velocity of the car change if it had (i) a tail wind of 10 m s^{-1}, (ii) a wind blowing at it westwards at 10 m s^{-1}, and (iii) a wind blowing at it westwards at 15 m s^{-1}?

(i) See Fig. 2.7.

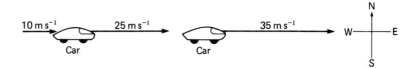

Figure 2.7

The resultant velocity $v = 25 + 10 = \underline{35\,\mathrm{m\,s^{-1}}\;\text{eastwards.}}$

(ii) See Fig. 2.8.

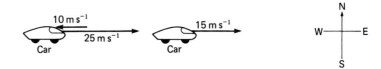

Figure 2.8

The resultant velocity $v = 25 - 10 = \underline{15\,\mathrm{m\,s^{-1}}\;\text{eastwards.}}$

(iii) See Fig. 2.9.

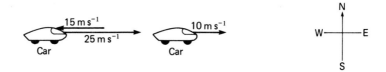

Figure 2.9

The resultant velocity $v = 25 - 15 = \underline{10\,\mathrm{m\,s^{-1}}\;\text{eastwards.}}$

The positive and negative signs in examples 2.4 and 2.5 for the addition of parallel vectors indicate vectors of opposite *directions*.

The addition of non-parallel vectors acting at a point is a little more tricky. Their different directions mean that they cannot add or subtract like ordinary numbers.

Two non-parallel vectors acting at a point are made to form half a parallelogram and the rest of the parallelogram is sketched in with dashed lines. The diagonal of the parallelogram then gives the **resultant**.

Example 2.6

Two footballers A and B kick a football with equal force simultaneously, but in different directions as shown in Fig. 2.10. What is the resultant force R on the ball?

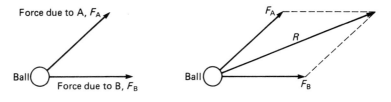

Figure 2.10

A scale has been used to draw the forces due to A and B. From this same scale the resultant R can be found by measuring the length of the diagonal of the parallelogram.

Example 2.7

A speedboat moves with a velocity of 30 m s^{-1} due east. What will the resultant velocity of the speedboat be if a wind blowing at 40 m s^{-1} due south acts upon it?

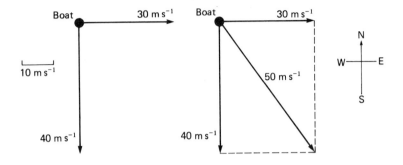

Figure 2.11

The scale in Fig. 2.11 is one cm to every 10 m s^{-1}. The diagonal of the parallelogram is 5 cm long, and so the resultant velocity of the speedboat is 50 m s^{-1}, in an approximately south-easterly direction.

The angle θ between the original velocity of the boat and its resultant velocity can be found by applying trigonometry (see Fig. 2.12).

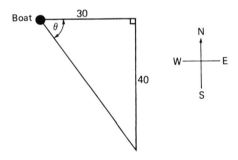

Figure 2.12

$$\tan \theta = \frac{30}{40} = 0.75$$

therefore

$$\theta = \underline{36°52'}$$

Thus, the bearing of the speedboat (the angle between north and the resultant velocity) is

$$\text{bearing} = 90° + 36°52' = \underline{126°52'}$$

Example 2.8

A piece of string with equal weights at each end is run across two pulleys. A third set of weights is attached at point B as shown in Fig. 2.13 such that all the weights are in equilibrium. If the tensions in the parts of the string AB and BC are 6 N each and the angle between them is 70°, find the tension T caused by the third set of weights.

Since all the weights are in equilibrium, the total force acting upwards at B must equal the downward force T acting at B. The resultant of the two

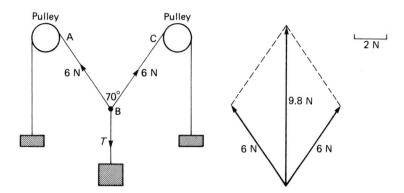

Figure 2.13

6 N tensions is found by measuring the length of the diagonal of the parallelogram and then applying the scale.

The length of the diagonal represents a force of 9.8 N upwards; thus,

$$T = \underline{9.8 \text{ N downwards}}$$

Example 2.9

Three different horses X, Y and Z pull the same load. If they suddenly bolt, each pulling the load with a force as shown in Fig. 2.14, find the resultant force R on the load.

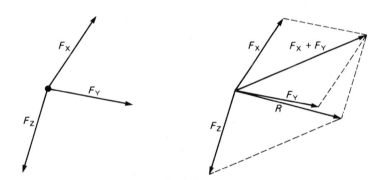

Figure 2.14

The resultant R is found by adding two of the forces and then adding the third one to their sum.

Subtraction of vector quantities

A negative vector has the opposite direction to the positive of that vector. Thus, subtraction of a vector from another is achieved by reversing its direction and then adding the two vectors.

Example 2.10

Two forces of magnitude 40 N and 50 N act at a point, at 50° to each other, as shown in Fig. 2.15. (i) Find their sum and (ii) subtract the smaller force from the larger.

(i) The sum is obtained by measuring the length of the diagonal of the parallelogram and then applying the scale.

From the diagram the length of the diagonal represents $\underline{82 \text{ N}}$ resultant force.

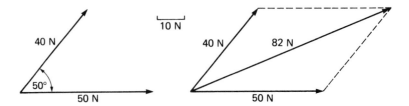

Figure 2.15

(ii) The difference is obtained by reversing the direction of the 40 N force and then measuring the length of the diagonal of the new parallelogram. The scale can then be once again applied. See Fig. 2.16.

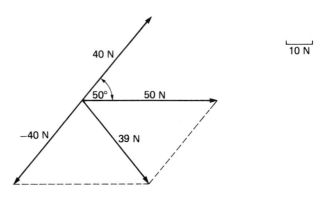

Figure 2.16

The difference is 39 N.

Resolution of vectors

This can be thought of as the reverse of vector addition. It is the splitting up of one vector into two vectors, which are called **components**. The easiest case to treat is the resolution of a vector into two perpendicular components.

The magnitude of a vector is represented by the length of the arrow. Thus, if the magnitude of a vector P is p it can be resolved into two perpendicular components R and Q whose magnitudes are $p \sin \theta$ and $p \cos \theta$ respectively. (See Fig. 2.17).

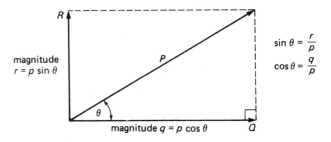

Figure 2.17

Example 2.11

A supersonic plane dives at an angle of 60° to the horizontal with a speed of 380 m s^{-1}. Find (i) the horizontal component of its velocity and (ii) the vertical component of its velocity.

(i) See Fig. 2.18. The magnitude of the horizontal component of the

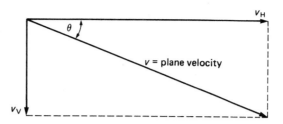

Figure 2.18

plane's velocity v_H is given by

$$v_H = v \cos \theta = 380 \times \cos 60° = \underline{190\ \mathrm{m\,s^{-1}}}$$

(ii) The magnitude of the vertical component of the plane's velocity v_V is given by

$$v_V = v \sin \theta = 380 \times \sin 60° = \underline{329\ \mathrm{m\,s^{-1}}}$$

Example 2.12

A system of three magnets as shown in Fig. 2.19 is used to keep a small steel pin at point X, by each applying a force of attraction upon it. If magnet C applies a force of 50 N, and the angles between C and B, and B and A, are as shown, find the forces applied by B and A. The resultant of two of any of the forces must be equal but opposite in direction to that of the third.

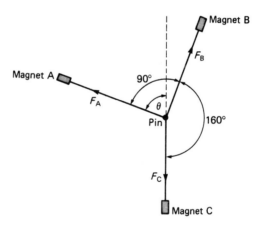

Figure 2.19

Since angle CB is 160° and angle BA is 90°, then $\theta = 70°$. So the force due to B is

$$F_B = F_C \sin \theta = 50 \times \sin 70° = \underline{47\ \mathrm{N}}$$

and the force due to A is

$$F_A = F_C \cos \theta = 50 \times 70° = \underline{17\ \mathrm{N}}$$

Mass of a body

Bodies vary in the way in which they oppose changes made in their motions. For example a cannon ball when fired will not be deflected by a human hand. It is said to have a *high inertia*. However, a small beach ball, when thrown, can be deflected by the human hand. It has a *small inertia*.

The numerical measure of inertia is called **mass**. The mass of a body is the *same anywhere*.

Types of force

Weight

The weight of a mass is the force of attraction of the Earth upon it.

$$\boxed{W = mg}$$ (unit: N)

In the outside world both mass and weight are thought to be the same, but from the Physics point of view they are *not*. As g varies geographically, so does W. This means weight is *not* the same everywhere.

Tension

This is the force occurring in a taut string, or a bicycle chain. The tension in the string is equal to the weight of any suspended body. It also occurs in a spring and for this reason a *spring balance* is graduated in newtons and *not* kg.

Thrust

The opposite of tension (or pull) is a thrust (or push).

Reaction

When two bodies are in contact, each exerts a thrust on the other and these thrusts are called **reactions**. As a thrust is a force and force has direction, a reaction whose direction is *perpendicular* to the surface of a body is called a **normal reaction**.

Friction

Frictional forces act along the surface between two bodies whenever one body tries to move over another. The direction of the friction *opposes* the relative motion of the bodies, i.e. the motion of one body as compared to that of the other.

The friction occurring when bodies are moving over each other is called **kinetic friction**. The friction occurring when a body simply rests on another is called **static friction**.

The friction due to a fluid, i.e. a liquid or a gas, is often called **drag** or the **viscous force** on the body in motion. The magnitude of the drag increases as the speed of the body increases. Another term in common use is **air resistance**, which is the drag on a body moving through air, due to the air.

Upthrust

This is the force a body experiences in an upward direction when it is partially or wholly immersed in a *fluid*, i.e. a liquid or a gas.

Centre of gravity

The force on a body due to the Earth's gravitational attraction (the **weight** of the body) is the sum of the forces of attraction on all the particles and atoms making up the body. The **centre of gravity** (CG) is the fixed point on or near to the body, through which the sum of these separate forces acts. As $W = mg$, it is also the point at which all the mass of the body appears to be concentrated. It is thus often called the **centre of mass** (CM).

For stability a body must preferably have a large base and a low centre of mass. A body suspended through its centre of mass is in **equilibrium**.

Generally the centre of mass of a symmetrical body made of material of uniform density is at the centre of the body.

Moments

The **moment** or **torque** of a force *about a point* is the product of the force and the *perpendicular* distance of its line of action from that point. It has units of N m. For example consider a trap door hinged at O, as shown in Fig. 2.20.

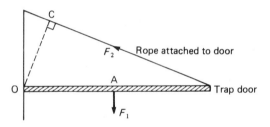

Figure 2.20

The moment of F_1 about O is $F_1 \times OA$ and the moment of F_2 about O is $F_2 \times OC$.

A **couple** consists of two equal but opposite parallel forces whose lines of action do not coincide. A moment produces a turning motion while a couple produces rotation. The line of action of a force is an imaginary line running through the arrow representing the force (see Fig. 2.21).

Figure 2.21

The forces, each of value F, used to open a tap are an example of a couple (Fig. 2.22). The *moment of a couple C* is the product of one of the forces and the perpendicular distance between the forces. For example, for the tap

$$C = (F \times OA) + (F \times OB) = F \times AB \qquad \text{(moment about O)}$$

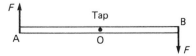

Figure 2.22

Principle of moments

When a body is in *equilibrium* the sum of the anticlockwise moments about any point is equal to the sum of the clockwise moments about the same point.

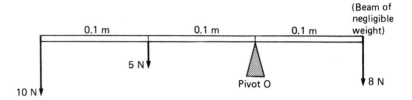

Figure 2.23

If the body is not in equilibrium, i.e. it is turning, there is an overall or **resultant** moment. In Fig. 2.23, the anticlockwise moments taken about O are

$$10 \times 0.2 \quad \text{and} \quad 5 \times 0.1$$

The clockwise moment taken about O is

$$8 \times 0.1$$

The sum of all the moments is

$$(10 \times 0.2) + (5 \times 0.1) - (8 \times 0.1) = +1.7 \, \text{N m}$$

A clockwise moment is given a negative sign, and therefore the resultant moment is an anticlockwise one. This means that the beam turns anticlockwise. If the resultant moment had been zero, the beam would have been in equilibrium.

Equilibrium

For a body to be in equilibrium, firstly it must satisfy the principle of moments and secondly the resultant force upon it must be zero.

Newton's laws of motion

First law

Every body continues in its state of rest or motion in a straight line with constant speed unless acted upon by an external force.

This law defines for us what a **force** is. It also tells us about the property called **inertia**. The inertia of a body is its reluctance to start moving or to stop moving. Inertia is measured in terms of mass. Inertial mass is the same everywhere and its value is decided by a comparison with a standard kilogramme mass.

Examples of this law include the motion of a rocket sent into deep space. The rocket will continue to move in a straight line at constant speed without the aid of engines, provided there are no external forces acting on it like, for example, the gravitational pull of a planet or star. A coin resting on a piece of card placed on top of the open mouth of a bottle will drop into the bottle if the card is sharply flicked away. A possible external force in this case is friction between the coin and the card but, as this acts over a very short period of time, its effect is negligible.

Second law

Before discussing this it is necessary to define the term 'linear momentum'.

Linear momentum is the product of the mass and the velocity of a body and is usually given the symbol p.

$$\boxed{p = mv}$$ (units: kg m s^{-1} or N s)

The student must take care to be able to distinguish between the terms **momentum** and **moment of a force**—they are *not* the same thing.

As velocity is a vector so is momentum.

Newton's second law tells us that the rate of change of momentum of a body is directly proportional to the resultant force acting on that body, and takes place in the direction of that force.

It is this law which gives rise to one of the most famous equations in Physics. From the law

$$F \propto \frac{\text{change in momentum}}{\text{change in time}}$$

and from the definition of momentum

$$F \propto \frac{\text{change in (mass} \times \text{velocity)}}{\text{change in time}}$$

But the mass of a body does not change in time, and so this can be taken out (this assumes that the body is not disintegrating while in motion).

$$F \propto \text{mass}\left(\frac{\text{change in velocity}}{\text{change in time}}\right)$$

and from the definition of acceleration

$$F \propto \text{mass} \times \text{acceleration}$$

In SI units this can be written as

$$\boxed{F = ma}$$

It is important to remember that as the law states, F is the *resultant* or *unbalanced* force acting on the body producing its motion.

This law, then, gives us a way of obtaining a value of the force we have defined in the first law.

Examples of Newton's second law

There are many examples of this law. A fast accelerating car will do considerably more damage in a collision with a wall than a slowly accelerating car. (The driver will continue to move forward if he is not wearing a seat belt, as predicted by Newton's first law of motion.) At a fair ground test-of-strength machine the higher the hammer is raised the greater its acceleration can be made and hence the greater the force with which it strikes the base-knob.

Exercise for the student: In what way is the packaging of eggs an example of Newton's second law of motion?

Students must be warned against giving the equation $F = ma$ as the law. It is *not* Newton's second law of motion. It is an equation which is a direct result of it.

Third law

To every action there is an equal but opposite reaction, where action and reaction are forces and act on different bodies.

This law is perhaps the most difficult of the three to understand. Surely if a person presses down on the Earth in order to jump, application of the law would mean that the Earth presents an equal but upward force upon that person and so he will not move? The important point is that the action and the reaction act on different bodies. The mass of the Earth is very large and so although it does experience a force and hence an acceleration, this is very small. The mass of the person is, compared to that of the Earth, much smaller and thus the acceleration which the person experiences is proportionally larger than that of the Earth.

Principle of conservation of linear momentum

When bodies in a system interact, there is no change in *total* linear momentum, provided no external force acts on the system.

An example of a system might be two particles or three cars on an isolated track. An interaction is often a collision. There are two types of collision, as follows.

Elastic collision
In this interaction the total *kinetic* energy before the collision is the same as the total kinetic energy after it for the system.

Inelastic collision
Here the total kinetic energy is less after the collision than before it. (Some of the initial kinetic energy gets transformed into other types of energy such as heat generated in a collision or the sound energy produced in a collision, or the work done in creating a dent, etc.)

Remember that linear momentum is conserved in *both types* of collision as long as no external force is applied.

Experimental investigations

Newton's laws and the principle of conservation of linear momentum can be investigated experimentally. The most common experiments involve *dynamics trolleys*. The dynamics trolley or trolleys are attached to ticker-tape which also passes through a timer. The ticker-timer makes 50 dots per second along the tape as it passes through it. (The timer works on mains a.c. and hence the number 50 since this is the frequency of mains a.c.). From the number of dots the time of motion can be found, and the *distance of travel* is obtained from the length of the tape which has passed through the timer.

In describing dynamics trolley experiments in exams it is always important to mention how the measurements of time and distance are obtained from the tapes.

Investigation of Newton's second law using dynamics trolleys

In these experiments the trolleys are placed on a ramp (Fig. 2.24). This is because any flat surface presents a friction force to bodies moving across it, thus affecting the bodies' acceleration.

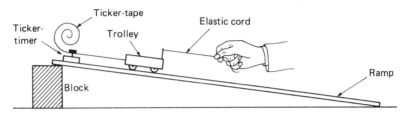

Figure 2.24

To prevent this the ramp is inclined until a trolley moving freely down it, moves with constant speed (i.e. the net force acting on the trolley is zero). This raising of the ramp is called **friction compensation**.

The relationship between *resultant force* and *acceleration* can be obtained by pulling a single trolley down the ramp with different numbers of elastic cords. If the length of the cords is kept constant (this takes a little practice), it is found that the acceleration doubles when two cords are used, trebles when three cords are used, and so on. Thus, the resultant force is *directly proportional* to the acceleration.

The acceleration of the trolley is found from the tape attached to it, by finding the velocity during, say, the first 0.2 s of the trolley's motion and then finding the velocity of the trolley during the final 0.2 s of the trolley's motion. The difference of these two values then gives the *change in velocity*, and this is divided by the *time interval* between the two.

It can also be shown that the acceleration of the trolley is *inversely proportional* to its mass. This is achieved by using only one elastic cord (and thus applying a constant force), attached to different numbers of trolleys stacked one upon the other. Doubling the mass of the trolleys halves the acceleration, tripling the mass of the trolleys reduces the acceleration to one third, and so on.

Investigation of momentum conservation

The conservation of linear momentum can also be observed using dynamics trolleys. In this case an interaction or collision is arranged. The easiest type of collision to perform is an *inelastic one*. For this a trolley with a pin sticking from the front of it is allowed to move freely down a *friction compensated* ramp. Halfway along this it collides with a stationary trolley which has a cork sticking out from its back. The two trolleys join and move off together down the ramp. The whole experiment is recorded on the ticker tape attached to the first trolley. The velocity before and after the collision is obtained from this tape and, either by finding the actual trolley masses or by treating them as 'unit' trolley masses, the *total momentum* before and after the collision can be calculated.

A *partially elastic* collision can be obtained by using a spring-loaded buffer instead of a pin in the case of the first trolley and an ordinary trolley in the case of the second stationary one. Both trolleys must be attached to ticker-tapes and the experiment is rather difficult to perform.

Examples of conservation of linear momentum

A collision is one type of interaction which follows the principle of conservation of linear momentum. Dynamics trolleys are used to investigate such events and the diagrams for an inelastic and partially elastic collision are shown in Figs 2.25 and 2.26. The term *partially* is used in the latter case because it is very difficult to produce a situation where the total kinetic energy before the collision is even approximately equal to the total kinetic energy after the collision.

Inelastic collision

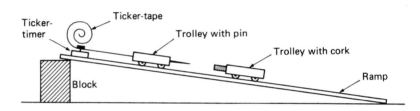

Figure 2.25

Partially elastic collision

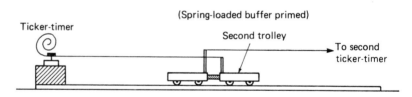

Figure 2.26

Example 2.13

Two bodies collide inelastically, one initially being at rest and the other moving at $25\,\mathrm{m\,s^{-1}}$. The one at rest has a mass of 2 kg and is twice as massive as the other. Both move off together after the collision, with what speed?

Applying the principle of conservation of linear momentum,

$$p_{\text{initial}} = p_{\text{final}}$$

so

$$m_1 v_1 + m_2 v_2 = (m_1 + m_2)v$$

and thus,

$$v = \frac{m_1 v_1 + m_2 v_2}{m_1 + m_2} = \frac{(1.0 \times 25) + (2.0 \times 0.0)}{1.0 + 2.0}$$

Therefore

$$v = \underline{8.3\,\mathrm{m\,s^{-1}}}$$

Since momentum is a vector, its direction needs to be considered during an elastic collision where the two bodies rebound off each other.

Example 2.14

If the moving mass in example 2.13 bounces directly back after collision, with speed u, and the body at rest is moved forward, with a speed of $10\,\text{m s}^{-1}$, find the value of u.

Let the 'forward' direction be positive, then

$$m_1 v_1 + m_2 v_2 = m_1(-u) + m_2 v_2^*$$

and so

$$-u = \frac{m_1 v_1 + m_2 v_2 - m_2 v_2^*}{m_1} = \frac{(1 \times 25) + (2 \times 0) - (2 \times 10)}{2}$$

Therefore

$$u = \underline{-2.5\,\text{m s}^{-1}}$$

Equilibrium

There are two general conditions which apply when a body is in equilibrium, as the result of a number of non-parallel forces acting in the *same plane* upon it.

The first condition is that the total or resultant force is *zero*. That is, the sum of the forces acting in one direction is equal to the sum of the forces acting in the opposite direction. The second condition concerns the principle of moments.

For example, a box of weight 40 N stands on a rough, flat surface (Fig. 2.27). A horizontal push of 20 N is applied to it, and it is seen to just move. The box is only just moving and so the total force on it is zero.

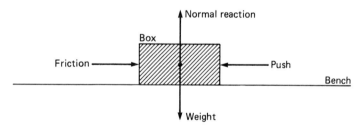

Figure 2.27

The weight of the box is balanced by an equal but opposite force—the normal reaction due to the bench. The friction is just about balanced by the push.

When the surface is *inclined* to the horizontal at angle θ (Fig. 2.28) then the force which acts perpendicular to the surface is no longer the weight of the box, but a component of it.

If the weight of the box is mg, then the component of the weight acting perpendicular to the surface is $mg\cos\theta$ and this is balanced by the equal but opposite normal reaction due to the surface. As the box does not slide down the surface, the other component of the weight $mg\sin\theta$ which is trying to pull the box down the incline must be balanced by the friction, which acts upwards along the slope against the direction of motion. Thus, the friction in this case is equal to $mg\sin\theta$.

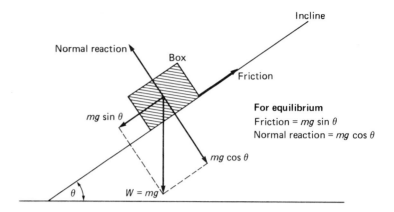

Figure 2.28

Archimedes' principle

The upthrust on an object partly or wholly immersed in a fluid is equal to the weight of the fluid displaced.

$$\text{upthrust } U = m_f g = \rho_f V_f g$$

where m_f, V_f and ρ_f are the mass, volume and density of the fluid displaced, respectively. (When the body floats, its weight must equal the upthrust. If it just floats and is almost totally submerged, the density of the body must therefore equal the density of the surrounding fluid—e.g. a floating balloon.)

Work and force

Work is done when a force moves the body to which it is applied in the direction of the force. It is a scalar quantity and is defined as the product of the applied force F and the distance moved by the point of application in the direction of the force s.

$$\boxed{W = Fs}$$

The **joule** (symbol **J**) is the unit of work. When work is done it usually accompanies a change or transformation of energy. For example, in the lifting of a mass using a rope and pulley, work is being done, and a conversion of chemical energy in muscles to gravitational potential energy occurs.

Work done by a tangential force producing rotation

A force F is applied to the circumference of a wheel, which is pivoted at its centre (Fig. 2.29). If the force moves the wheel a distance s then the work done by the force is

$$W = Fs$$

$$\boxed{W = Fr\theta} \quad (\theta \text{ in radians})$$

where Fr is the moment of F about O.

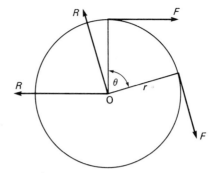

Figure 2.29

Since the point O remains fixed, there is a reaction R there and so F and R form a couple, whose moment is also Fr.

$$\text{work done by a couple} = \text{moment of couple} \times \text{angle turned (in radians)}$$

Energy

This is the property of a system that is a measure of its capacity for doing work. The unit of energy is also the **joule.**

Einstein's energy–mass relationship

Matter is a form of energy and thus, for an isolated system, when a change in mass occurs, a corresponding change in the energy of the system also occurs. An example of this is the energy liberated in a nuclear reaction. If the change in mass is Δm then the corresponding change in energy will be given by

$$\Delta E = \Delta mc^2 \qquad (c = \text{speed of light})$$

Power

Power is the rate of doing work and the unit of power is the **watt** (symbol **W**).

$$\text{power} = \frac{\text{work done } W}{\text{time taken } t}$$

But
$$W = Fs$$
Thus,
$$\text{power} = Fs/t$$
But
$$s/t = \text{speed}$$
Therefore
$$\text{power} = \text{force} \times \text{speed}$$

(Occasionally power is written as energy transferred/time.) The watt can thus be expressed also as $\text{W} = \text{J s}^{-1}$.

Potential energy (PE)

This is the energy a system possesses because of the *position* of its relative parts. For example two magnets with N-poles very close together have a greater amount of PE than when the magnets are widely separated. The most common type of PE is **gravitational potential energy**. There is no general formula for PE. When a mass m is raised through a distance h, the work done is mgh. The mass has now the ability or capacity by virtue of its position to do work mgh if it is dropped.

$$\boxed{\text{gravitational PE} = mgh}$$

Kinetic energy (KE)

This is the energy possessed by a moving body. The equation for KE is

$$\boxed{\text{KE} = \tfrac{1}{2}mv^2}$$

Transformation of energy

Consider the following example.

When a ball is thrown into the air, a certain amount of chemical energy in the muscles is used up. This energy does not just disappear but is transformed into other types of energy as the ball continues its motion.

As the ball rises it gets slower, so it loses KE, but since it is getting higher it gains gravitational PE. At the highest point in its motion all the energy it has is in the form of gravitational PE and this should equal exactly the amount of chemical energy previously mentioned.

On the way down the ball loses gravitational PE but gains KE. Finally when the ball hits the ground all the KE is converted into sound, heat and work done, perhaps in making a dent in the ground.

Law of conservation of energy

Energy cannot be created or destroyed, only converted from one form to another.

When work is done on a system the energy converted is equal to the work done on the system. For example:

$$Fs = \tfrac{1}{2}\Delta mv^2$$

Uniform motion in a circle

In Physics, angles are expressed in **radians** rather than in degrees. The angle θ in radians is the ratio of the length of the arc subtended at the centre of a circle by angle θ to the radius of the circle (see Fig. 2.30).

$$\theta \text{ (in radians)} = s/r$$

Thus,

$$\boxed{s = r\theta}$$

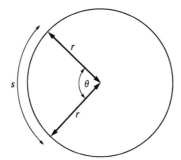

Figure 2.30

Angular velocity

When a particle moves in a circle, around a centre O, the angle through which it goes, θ, changes with time. The rate of change of θ with time is called the **angular velocity**, ω. If ω is constant, then

$$\omega = \frac{d\theta}{dt} = \frac{\theta}{t} \qquad \text{(units: rad s}^{-1}\text{)}$$

The particle travels a distance s in time t and so its *linear velocity* has the magnitude

$$v = s/t = r\theta/t \qquad \text{(units: m s}^{-1}\text{)}$$

Therefore

$$v = r\omega$$

Now $t = s/v$ and if the particle completes one circle then $s = 2\pi r$ and so the time taken T for this is given by

$$T = 2\pi r/v = 2\pi/\omega$$

T is called the **period of rotation** and it is the time taken for a body to revolve once around a fixed point. Its unit is thus the second.

The **frequency of rotation** f is the number of rotations a body makes around a fixed point per second.

$$f = 1/T = \omega/2\pi \qquad \text{(unit: s}^{-1}\text{)}$$

(The s^{-1} is called the **hertz** and has the symbol Hz.)

Acceleration of a body in circular motion

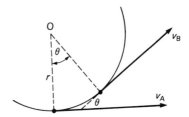

Figure 2.31

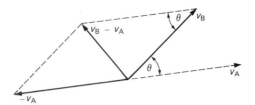

Figure 2.32

If the vectors v_A and v_B represent the velocities of a body at two different points during its circular motion (Fig. 2.31), then

$$\text{acceleration} = \frac{\text{change in velocity}}{\text{time}} = (v_B - v_A)/t$$

In order to perform the subtraction of the two velocities it is necessary to apply the *parallelogram of vectors* (Fig. 2.32).

If the body moves through an angle θ around the centre of the circle O, with a constant speed v, the trigonometry of the parallelogram of vectors is such as to make the change in velocity $v_B - v_A$ have magnitude $v\theta$ where θ is in radians as usual.

Magnitude and direction of the acceleration of a body in circular motion

Having applied the parallelogram of vectors, the following expressions for acceleration can now be written

$$\text{acceleration} = v\theta/t = v\omega$$

$$= v^2/r = \omega^2 r \qquad (\text{since } v = r\omega)$$

Acceleration is a vector and the direction of the acceleration of a body in circular motion is always *towards the centre* of the circular path. This is indicated by the insertion of a negative sign.

$$\boxed{a = -\omega^2 r} \qquad (\text{units: m s}^{-2})$$

Centripetal force

This is the *inward radial* force imposed by the constraining system needed to keep a body moving in a circle.

$$F = ma = mv^2/r = mr\omega^2$$

The centrifugal force is an equal but opposite force acting on the body moving in the circle, i.e. an outward force.

Examples of centripetal force include: the gravitational attraction between the Sun and the planets, the motion of a car rounding a bend, and the tension in a rope attached to a stone when the stone is being whirled around.

Centrifuge

This separates particles from a suspension. Balanced tubes with the liquid rotate rapidly about a central point. The tubes are free to slant away from this point as they move around.

A drop of liquid in the tube experiences a force $mr\omega^2$ from surrounding drops just to keep it in motion. If the drop is replaced by a solid particle of mass m^* it will still experience a force $mr\omega^2$. But, if $m^* > m$ then $mr\omega^2$ is not enough to keep the solid particle in motion, so it moves outwards, to the bottom of the tube. The bottom of the tube can supply the extra force needed because of the *larger value* of r.

Ultracentrifuges used for studying polymers (other applications include conductor study, space research, and medicine) have reached rotational frequencies of 25 000 Hz.

Kepler's laws of planetary motion

(1)   The planets describe ellipses using the Sun as one of their foci.
(2)   A straight line joining the Sun and a planet sweeps out equal areas in equal times.
(3)   The squares of the planetary periods of revolution are proportional to the cubes of their average distances from the Sun.

$$\boxed{T^2 \propto r^3}$$

These are *empirical laws*, i.e. based on observation and not on mathematical derivation. They give the position of the planets at any time.

Example 2.15

A model radio-controlled boat of mass 2 kg is made to move in a circle of radius 4 m with a uniform speed of $6 \, \mathrm{m \, s^{-1}}$. Find (i) its angular velocity and (ii) the centripetal force on it.

(i) $v = r\omega$

and $\omega = \dfrac{v}{r} = \dfrac{6}{4} = \underline{1.5 \, \mathrm{rad \, s^{-1}}}$

(ii)   Centripetal force

$$F = mr\omega^2$$

Thus,
$$F = 2 \times 4 \times 1.5^2 = \underline{18 \, \mathrm{N}}$$

Example 2.16

The metal arm of a fairground ride propels a toy rocket of mass 140 kg in a horizontal circle of radius 5 m with a constant speed of $8 \, \mathrm{m \, s^{-1}}$. If the arm is slightly inclined to the horizontal and $g = 9.8 \, \mathrm{m \, s^{-2}}$, find the tension in the arm.

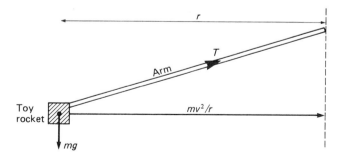

Figure 2.33

Applying the parallelogram of vectors to Fig. 2.33, then for equilibrium

$$T^2 = \left(\frac{mv^2}{r}\right)^2 + (mg)^2 \qquad \text{(Pythagoras)}$$

Then $T^2 = \left(\frac{140 \times 8^2}{5}\right)^2 + (140 \times 9.8)^2$

and $T = \underline{2.3 \times 10^3 \text{ N}}$

Example 2.17

A bicycle and rider of total mass 70 kg travel round a circular banked course with a uniform speed of 15 m s^{-1}. If the reaction at the bicycle wheels is normal to the track and the horizontal radius of the course is 50 m, find the inclination of banked track and the reaction.

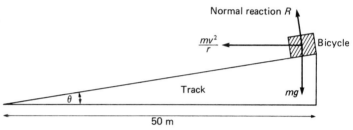

Figure 2.34

Applying the parallelogram of vectors to Fig. 2.34, then for equilibrium

$$R\cos\theta = mg \quad \text{and} \quad R\sin\theta = mv^2/r$$

Thus,

$$\frac{mg}{\cos\theta} = \frac{mv^2}{r\sin\theta} \quad \text{and} \quad \tan\theta = \frac{v^2}{rg} = \frac{15^2}{500}$$

Therefore

$$\theta = \underline{24°}$$

From $R = \dfrac{mg}{\cos\theta}$

$$R = \frac{(70 \times 10)}{\cos 24°} = \underline{766 \text{ N}}$$

Newton's law of universal gravitation

Every particle of matter in the universe attracts every other particle with a force which is directly proportional to the product of their masses and inversely proportional to the square of their distance apart.

$$F \propto \frac{m_1 m_2}{r^2}$$

which is written as

$$\boxed{F = \frac{Gm_1 m_2}{r^2}}$$

The constant of proportionality G is called the **universal gravitational constant**.

The value of G is taken as 6.664×10^{-11} N m^2 kg^{-2}.

Testing gravitation

For a planet of mass m and distance r from the Sun of mass M,

$$F = \frac{GMm}{r^2} = \text{centripetal force on the planet}$$

Thus,

$$\frac{GMm}{r^2} = \frac{mv^2}{r}$$

and so

$$v^2 = \frac{GM}{r}$$

Now recalling that the period of rotation for a body moving in a circle is $T = 2\pi r/v$, then

$$\frac{GM}{r} = \left(\frac{2\pi r}{T}\right)^2$$

and rearranging this gives

$$T^2 = \frac{4\pi^2 r^3}{GM}$$

which is of the form

$$T^2 \propto r^3$$

and this is **Kepler's third law** of planetary motion. The effects of gravitation between planets led to the prediction of two new planets—Neptune in 1846 and Pluto in 1930.

Mass of the Earth, M_e

We assume the Earth to be spherical and its centre of gravity to be at its centre. A mass m on the Earth experiences a force

$$F = \frac{GM_e m}{r^2} = mg$$

where r is the radius of the Earth.

Therefore, rearranging, we have

$$M_e = \frac{r^2 g}{G}$$

and, by introducing known values into this equation, M_e is found to be about 6.0×10^{24} kg.

Mass of the Sun, M

Now, if T is the period of rotation of the Earth around the Sun, then

$$T = 365 \times 24 \times 60 \times 60 = 3.2 \times 10^7 \text{ s}$$

By applying the previously derived expression for T,

$$T^2 = \frac{4\pi^2 r^3}{GM}$$

where M is the mass of the Sun and r is the Sun–Earth distance, then M is found to be about 2.0×10^{30} kg.

Parking orbit

These are orbits in which a satellite remains over the same spot on the Earth all the time. If the orbit is a distance R from the centre of the Earth,

$$\frac{mv^2}{R} = \frac{GMm}{R^2}$$

where M is the mass of the Earth and m is the mass of the satellite. But

$$GM = r^2 g$$

where r is the radius of the Earth. Therefore

$$v^2 = \frac{gr^2}{R}$$

The period of rotation of the satellite is

$$T = \frac{2\pi R}{v}$$

and so this can be rewritten as

$$T^2 = \frac{4\pi^2 R^3}{gr}$$

Weightlessness

A body in orbit is a freely falling one. Free fall is unimpeded fall in a gravitational field (g is now usually called the acceleration of free fall). If the body is a spacecraft, the free fall acceleration g^* of a particle inside *relative* to the spacecraft is

$$g^* = g - a$$

where a is the acceleration of the particle inside and g is the acceleration of the spacecraft itself. But $g = a$ and so $g^* = 0$.

The weight of the particle *relative* to the spacecraft is

$$\boxed{W^* = mg^* = 0}$$

Magnitude of g

Outside the surface of the Earth

$$F = mg = \frac{GM_e m}{r^2}$$

therefore

$$g = \frac{GM_e}{r^2}$$

This last equation is of the form $y = a/x^2$ where a is a constant (as x increases y decreases).

Inside the Earth

$$g = \frac{GM_e}{r^2} \qquad \text{(again)}$$

But the deeper into the Earth we go, the smaller M_e (which is the mass of the Earth that is exerting a force) gets. M_e is *not constant*.

$$M_e = \rho \tfrac{4}{3}\pi r^3 \qquad \text{(assuming a spherical Earth)}$$

therefore

$$g = \frac{G\rho \tfrac{4}{3}\pi r^3}{r^2}$$

and so

$$g = G\rho \tfrac{4}{3}\pi r$$

As long as the mean density of the Earth, ρ, is constant, this equation is of the form $y = mx$.

Variation of *g*

The variation of g from the centre of the Earth is illustrated in Fig. 2.35.

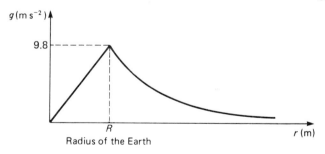

Figure 2.35

The value of g also varies *geographically* because:

(1) The Earth is an oblate spheroid and not a sphere.

$$mg = \frac{GM_e m}{r^2}$$

therefore

$$g = \frac{GM_e}{r^2}$$

But r is greater at the equator than at the poles (Fig. 2.36), so g is smaller at the equator than at the poles.

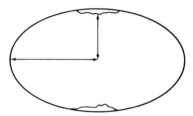

Figure 2.36

(2)   Rotation of the Earth.
     A body on the surface of the Earth, apart from at the poles, experiences a centripetal force due to the rotation of the Earth (Fig. 2.37). This force is supplied by part of the Earth's gravitational field.

$$\text{centripetal force} = mg - mg_0$$

$$\cos \theta \, \frac{mv^2}{R} = mg - mg_0$$

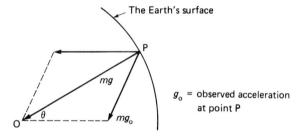

Figure 2.37

g_0 = observed acceleration at point P

(3)   Local deposits of heavy or light ores.
(4)   Sideways attraction of mountains, etc.

Methods for finding g

(1)   Free fall of mass through ticker-tape timer.
(2)   Electrical timing of freely falling mass.
(3)   Simple pendulum, compound pendulum, Kater's pendulum.
(4)   Oscillating loaded spring.

For the following examples assume $G = 6.7 \times 10^{-11} \, \text{N m}^2 \, \text{kg}^{-2}$.

Example 2.18

Using Newton's law of gravitation, find the mean density of a planet whose radius is 5500 km and which has an acceleration due to gravity of $7 \, \text{m s}^{-2}$.

From Newton's law, if the planet has mass M and radius r, then a mass m at its surface experiences a force of

$$F = ma = \frac{GMm}{r^2}$$

and therefore

$$M = \frac{ar^2}{G}$$

If the planet is spherical and has density ρ, then

$$M = \frac{4}{3}\pi r^3 \rho = \frac{ar^2}{G}$$

Thus,

$$\rho = \frac{3a}{4\pi rG} = \frac{3 \times 7}{4\pi \times 5500 \times 10^3 \times 6.7 \times 10^{-11}} = \underline{4500\,\text{kg}\,\text{m}^{-3}}$$

Example 2.19

A shuttle-type spacecraft is to remain in parking orbit over the Earth's equator. Find the orbit radius and the linear velocity of the craft if the mass of the Earth $M = 5 \times 10^{24}\,\text{kg}$.

If m and r are the craft mass and orbit radius respectively, then the force on the craft is

$$F = mr\omega^2 = \frac{GMm}{r^2}$$

and thus,

$$r^3 = \frac{GM}{\omega^2} = \frac{GMT^2}{4\pi^2}$$

$$r = \left(\frac{GMT^2}{4\pi^2}\right)^{1/3} = \left(\frac{6.7 \times 10^{-11} \times 5 \times 10^{24} \times (3600 \times 24)^2}{4\pi^2}\right)^{1/3}$$

Thus,

$$r = \underline{3.99 \times 10^7\,\text{m}} \qquad (T = \text{period of orbit} = 1\text{ day})$$

$$v = r\omega = \frac{2\pi r}{T} = \frac{2\pi \times 3.99 \times 10^7}{3600 \times 24} = \underline{2.9 \times 10^3\,\text{m}\,\text{s}^{-1}}$$

Example 2.20

The Sun is a star in a group called the Milky Way galaxy (see Fig. 2.38). The Sun is 4.6×10^{20} m from the galactic centre and orbits it with a speed $v = 3 \times 10^5\,\text{m}\,\text{s}^{-1}$. Find the mass M of the galaxy.

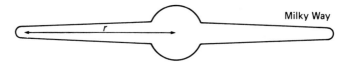

Milky Way

Figure 2.38

If the Sun has mass m and is distance r from the centre of the galaxy, then from Newton's law

$$\frac{GMm}{r^2} = \frac{mv^2}{r}$$

and so

$$M = \frac{v^2 r}{G}$$

Then

$$M = \frac{(3 \times 10^5)^2 \times 4.6 \times 10^{20}}{6.7 \times 10^{-11}} = \underline{6.2 \times 10^{41} \, \text{kg}}$$

The mass outside the radius r is not included.

Terminal speed

Viscosity is a measure of the drag of a fluid on a body. A sphere falling vertically through a viscous fluid experiences:

(1)   its weight W downwards,
(2)   upthrust U (from Archimedes' principle) upwards equal to the weight of fluid displaced,
(3)   viscous resistance F upwards (increases with speed).

The *resultant* downward force is then $W - U - F$. The sphere accelerates until F has reached a value where $W - U - F = 0$. Now there is no resultant force acting on the sphere and it continues downwards with a constant velocity called the **terminal speed**.

Gravitational potential, *U*

The potential U at a point due to the gravitational field of the Earth is equal to the work done in taking a *unit mass* from infinity to that point. This is similar to electric potential. The potential at infinity is usually taken as *zero*.

The work done by gravity in moving a *unit mass* a small distance δr towards the Earth is

$$\delta W = \text{force} \times \text{distance moved in direction of force}$$

$$\delta W = \frac{GM_e}{r^2} \delta r$$

The potential at a point of distance a from the Earth's centre is then a sum of these little bits of work.

$$\boxed{U_a = \frac{-GM_e}{a}}$$

The negative sign indicates that the potential at infinity is higher than the potential close to the Earth.

Rotational motion of a rigid body

For a rigid body rotating about a fixed axis O, the angular velocity $d\theta/dt$ or ω is the same for every particle.

$$\text{KE of a particle} = \tfrac{1}{2} m_p v_p^2$$
$$= \tfrac{1}{2} m_p r_p^2 \omega^2$$

where r_p is the perpendicular distance of the particle from the axis O.
Therefore the total KE of the body is

$$\tfrac{1}{2} m_1 r_1^2 \omega^2 + \tfrac{1}{2} m_2 r_2^2 \omega^2 + \ldots = \tfrac{1}{2} \omega^2 (\Sigma \, mr^2)$$

The quantity $\Sigma\, mr^2$ is called the **moment of inertia**, I, of the body about the axis O:

$$I = \Sigma\, mr^2 \qquad KE = \tfrac{1}{2}I\omega^2$$

The units of moment of inertia are $kg\, m^2$.

Examples of moment of inertia

Body	Moment of inertia ($kg\, m^2$)
Uniform rod of length l	$Ml^2/12$ about centre
	$Ml^2/3$ about one end
Ring of radius R	MR^2 about centre
Circular disc of radius r	$Mr^2/2$ about centre
Solid cylinder of radius q	$Mq^2/2$ about axis of symmetry
Sphere of radius a	$2Ma^2/5$ about axis through centre

Kinetic energy of a rolling body

When a body rolls down a plane it is rotating as well as moving bodily along. Therefore

$$\text{total KE of body} = \tfrac{1}{2}Mv^2 + \tfrac{1}{2}I\omega^2$$

Now if the body has an initial velocity $u = 0\,m\,s^{-1}$, then

$$v^2 = 2as$$

where s is the distance travelled. Also

$$v = r\omega$$

where r is the radius of the rotating body. So

$$\text{total KE of body} = \tfrac{1}{2}M2as + \tfrac{1}{2}\frac{Iv^2}{r^2} = Mas + \frac{Ias}{r^2}$$

Rearranging, we obtain:

$$a = \frac{\text{total KE}}{Ms + Is/r^2}$$

This means that the acceleration of the cylinder is *inversely proportional* to the moment of inertia of the cylinder.

Example
Two cylinders A and B roll down a plane without slipping. If both have equal masses but A has a smaller moment of inertia than B, A will roll down faster (e.g. B might be hollow.)

Angular momentum, *L*

This is the product of moment of inertia and angular velocity, and is a vector quantity.

$$\boxed{L = I\omega}$$ (units: $kg\, m^2\, rad\, s^{-1}$)

Conservation of *L*

Angular momentum about an axis of a given rotating body or system of bodies is constant if no external couple acts about that axis.

For example a diver will curl up in a dive to make *I* as low as possible, but since *L* cannot change ω increases. So the diver is able to perform more somersaults.

A similar case occurs when a skater spins on one skate. In order to spin faster the skater brings his arms in. (This reduces *I* but, since angular momentum must be conserved, his angular velocity increases.)

Example 2.21

A metal disc rolls along a flat surface with a speed of $0.6 \, \mathrm{m \, s^{-1}}$. Find its total kinetic energy if it does not slip and it has a mass of $0.3 \, \mathrm{kg}$.

For a disc of mass *M* and radius *r*, the moment of inertia about its centre is

$$I = Mr^2/2$$

and

$$v = r\omega$$

Thus, total kinetic energy *E* is

$$E = \tfrac{1}{2}Mv^2 + \tfrac{1}{2}I\omega^2 = \tfrac{3}{4}Mv^2$$

Therefore

$$E = 0.75 \times 0.3 \times 0.6^2 = \underline{0.08 \, \mathrm{J}}$$

Example 2.22

A wooden log rolls down a river bank without sliding. If the bank is inclined to the horizontal at 34° and the log is uniform throughout, find the time it takes to move 3 m, from rest. (Assume $g = 10 \, \mathrm{m \, s^{-2}}$.)

If the log has mass *M* and radius *r*, then the moment of inertia about its axis is

$$I = \frac{Mr^2}{2}$$

and the total kinetic energy, as in example 2.21, is

$$E = \underline{\tfrac{3}{4}Mv^2}$$

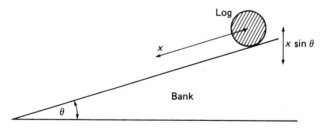

Log

x

$x \sin \theta$

Bank

θ

Figure 2.39

see Fig. 2.39. Now, applying

$$s = (u+v)t/2$$

then

$$v = 2x/t$$

since $u = 0 \, \mathrm{m \, s^{-1}}$.

From the *conservation of energy*, gain in KE = loss in PE. Thus,

$$\tfrac{3}{4}M\left(\frac{2x}{t}\right)^2 = Mgx\sin\theta$$

Therefore

$$t = \sqrt{3x/g\sin\theta} = \sqrt{9/(10\times\sin 34°)}$$

and

$$t = \underline{1.3\,\text{s}}$$

Example 2.23

A satellite spinning in space has moment of inertia about its axis of rotation of $37\,\text{kg}\,\text{m}^2$, and one of $53\,\text{kg}\,\text{m}^2$ when it spins with the panels carrying its solar cells stretched out. Find the ratio of the angular velocities when the panels are folded to when they are open.

(Use the subscripts f and o for folded and open panels respectively.)
From the *conservation of angular momentum*,

$$L_f = L_o$$

and so

$$I_f\,\omega_f = I_o\,\omega_o$$

Thus,

$$\frac{\omega_f}{\omega_o} = \frac{I_o}{I_f} = \frac{53}{37} = \underline{1.4}$$

Hence the satellite spins *faster* when the panels are folded.

Simple harmonic motion (shm)

A body which performs **simple harmonic motion** has an acceleration which is directly proportional to its displacement from a fixed point and which is directed towards that fixed point. This can be written as

$$\boxed{a \propto -x \quad\text{or}\quad a = -\,constant\cdot x}$$

(The negative sign indicates that the acceleration is directed towards a particular fixed point.) To illustrate shm we shall consider the following case.

A point P moves in a circle with *uniform* angular velocity ω (Fig. 2.40). Then its speed is given by $v = r\omega$.

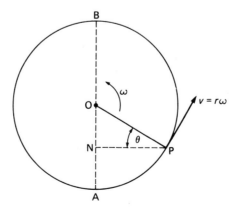

Figure 2.40

The acceleration of N is then a component of this.

$$a = -\omega^2 r \sin\theta$$

From Fig. 2.40, the displacement of N is then

$$x = r\sin\theta$$

Therefore the acceleration of N can be written as

$$a = -\omega^2 x$$

Since the angular velocity is a constant, N *is performing shm.*

When P goes around once, then N moves up and down once. So the period of oscillation of N is the same as the period of rotation of P. Therefore

$$\text{period of oscillation of N is } T = \frac{2\pi}{\omega}$$

The **amplitude** of the oscillation is defined as the maximum displacement from equilibrium. (For N the amplitude of oscillation is r.)

The **velocity** of N is the component of P's velocity parallel to the line AB. The velocity of P can be resolved into two components using the parallelogram rule for vectors. The component parallel to the line **AB** is $\omega r\cos\theta$. Now as θ is related to angular velocity by the expression $\theta = \omega t$, the velocity of N can be written as

$$\boxed{v = \omega r\cos\omega t}$$

By applying trigonometry, another expression for the velocity of N can be found.

$$\cos^2\theta = 1 - \sin^2\theta$$

therefore

$$\boxed{v = \omega r\sqrt{1 - \sin^2\theta}}$$

But

$$\sin\theta = x/r$$

therefore

$$v = \omega r\sqrt{1 - (x/r)^2}$$

and so

$$v = \omega\sqrt{r^2 - x^2}$$

Generally, for any body performing shm with amplitude of oscillation A, the velocity can thus be written as

$$\boxed{v = \omega\sqrt{A^2 - x^2}}$$

Graphical treatment of shm

The graphs in Fig. 2.41 have sinusoidal form but are out of step with each other. There is a difference of $T/4$ between graphs I and II and $T/4$ between graphs II and III. This is called **phase difference**. Since the graphs are sinusoidal we can talk about the phase difference in terms of

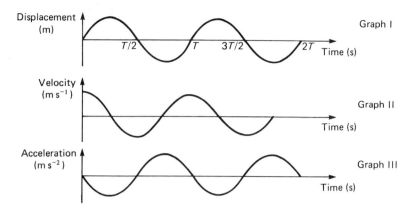

Figure 2.41

degrees or radians. Graphs I and II have a phase difference of 90° or $\pi/2$ radians (similarly for graphs II and III).

We say graph I is out of phase with graph II by $\pi/2$. This angle is called the **phase angle** (given the symbol ϕ).

The **phase angle** is the angle between vectors representing two harmonically varying quantities that have the same period.

An example is given in Fig. 2.42.

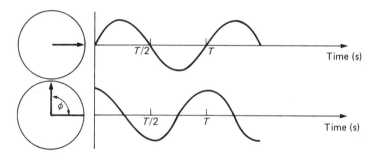

Figure 2.42

Note: As shm is sinusoidal, sine can be replaced by cosine and vice versa for the formulae given.

Harmonically varying quantities include the displacement of a pendulum from equilibrium, displacement of a weighted spring from equilibrium.

Experiments on the simple pendulum

By measuring the average periods of oscillation for a pendulum of varying length, then varying mass and then varying amplitude, it is found that the period is only affected by the change in the length of the pendulum string.

Frequency of oscillation

For a vibrating system, the frequency of oscillation is the number of oscillations performed in one second. It is equal to the reciprocal of the period of oscillation and has the unit the **hertz** (symbol Hz). (Refer also to the section on circular motion.)

$$\text{frequency of oscillation} = \frac{\text{number of oscillations}}{\text{time of oscillation}}$$

Simple pendulum

This is a small bob of mass m suspended by a light, inextensible thread of length L from a fixed point, as illustrated in Fig. 2.43.

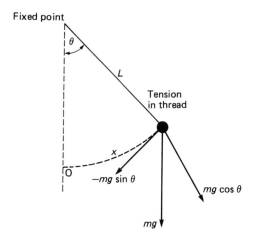

Figure 2.43

The minus sign indicates that the force component is *towards* O. The *unbalanced force* on the bob causing it to accelerate is

$$F = ma = -mg \sin\theta$$

But, for small θ,

$$\sin\theta = \theta \qquad (\theta \text{ in radians})$$

Therefore

$$ma = -mg\theta$$

But the displacement $x = L\theta$, so

$$ma = -mg\frac{x}{L}$$

$$a = -g\frac{x}{L}$$

This is of the form $a = -\text{constant} \cdot x = -\omega^2 x$, therefore the motion is shm. The period is

$$T = \frac{2\pi}{\omega} = 2\pi\sqrt{\frac{L}{g}}$$

Determination of *g*

Rearranging the above equation gives

$$L = \frac{g}{4\pi^2} T^2$$

This is of the form $y = mx$. If an experiment is carried out to measure the *average* period of oscillation T for different lengths of pendulum L and the results are plotted on a graph of L versus T^2, a *straight line* passing through the origin should be obtained if the pendulum is *executing shm*.

The slope of the straight line graph will be $g/4\pi^2$, from which a value of g can be found.

Energy exchanges in shm

(1) Centre of oscillation: KE = max.   PE = 0
(2) End of oscillation: KE = 0   PE = max.

Energy and shm

From the principle of conservation of energy,

$$\text{total energy in shm} = \text{max. KE} = \tfrac{1}{2}mr^2\omega^2$$
$$\text{(or max. PE)}$$

where r is the amplitude of oscillation.

Spiral spring or thread

Hooke's law

For a range of tensions in a wire, the extension produced is directly proportional to the force causing it. The range lies within the **elastic limit**. This means

$$\boxed{T = ke}$$

where k, the constant of proportionality, is called the **spring constant**.

If a mass m is put on a spring it stretches by a length e. From Hooke's law then

$$\boxed{mg = ke}$$

When the mass vibrates and goes a further distance down, x, the new tension is $k(e + x)$; see Fig. 2.44.

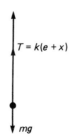

$T = k(e + x)$

mg

Figure 2.44

So the net upward force is

$$F = -[k(e+x)-mg] = ma$$

Multiplying out the brackets:

$$ma = -ke + mg - kx$$

But we already have $mg = ke$ from above, and therefore $-kx = ma$.

$$a = \frac{-kx}{m}$$

Again this is *shm* and the angular velocity is $\sqrt{k/m}$.

Potential energy of spring

The potential energy of a spring is $\frac{1}{2}kx^2$.

Potential energy of pendulum

The potential energy of a pendulum is

$$mgh = mgL(1-\cos\theta)$$

Damped oscillations

The oscillations of a pendulum or a mass on a spring eventually die away, due to air resistance and friction. This is called **damped motion**. Anything which is undamped is a **free oscillator**.

Example 2.24

A wine bottle with a long neck floats vertically in a pool of water. If it is pushed down a short distance x and then released, show that it will execute shm. Explain what occurs if the amplitude of oscillation is such that the main part of the bottle along with the neck partially rises above the water surface.

Figure 2.45

See Fig. 2.45. If the bottle has mass m and the main part of it has cross-sectional area A, and the density of the water is ρ, then the net upward force (from Archimedes' principle) is

$$F = -m_w g \qquad (m_w = \text{mass of water displaced})$$

Thus,

$$F = -A\rho xg$$

As a result of the force the bottle suffers an acceleration a, which is given by

$$F = ma = -A\rho gx$$

and therefore

$$a = -\frac{A\rho gx}{m}$$

This is now of the form $a \propto -x$ and thus the bottle is in shm.

If the main part of the bottle rises out of the water during its oscillation, then the cross-sectional area A will no longer be a constant. The bottle will then not perform shm.

Example 2.25

The piston movement of a car engine is almost shm. If the piston makes 9 strokes per second, each stroke being 10 cm and equal to twice the amplitude, find (i) the period of oscillation and (ii) the velocity and acceleration both at the mid-point of the stroke and at the amplitude of the oscillation.

(i)   Period $T = 1/f = 1/9 = \underline{0.1\,s}$.

(ii)   For shm

$$a = -\omega^2 x = -4\pi^2 f^2 x$$

At the mid-point of the stroke

$$x = 0.0\,\text{m}$$

thus,

$$a = \underline{0\,\text{m}\,\text{s}^{-2}}$$

For shm the velocity

$$v = \omega \sqrt{A^2 - x^2} = 2\pi f \sqrt{A^2 - x^2}$$

At the mid-point of the stroke, $x = 0.0$ m, therefore

$$v = 2\pi f \sqrt{A^2} = 2\pi \times 9 \times 0.05$$

Thus,

$$v = \underline{2.8\,\text{m}\,\text{s}^{-1}}$$

At the amplitude,

$$a = -\omega^2 A = -4\pi^2 \times 9^2 \times 0.05 = \underline{-160\,\text{m}\,\text{s}^{-2}}$$

At the amplitude,

$$v = \omega \sqrt{A^2 - A^2} = \underline{0.0\,\text{m}\,\text{s}^{-1}}$$

The point to note from example 2.25 is that the velocity of a body is zero when it is performing shm at the point of greatest displacement, while the acceleration there is a maximum, and vice versa for the body passing through the zero position. These two situations are often confused by students.

Example 2.26

A laboratory model of a gas consists of a flat, circular plate which oscillates up and down in a cylindrical plastic tube. Small beads placed on top of the plate are meant to represent gas molecules. If the plate performs shm, with a period of 0.3 s, find the maximum amplitude of the plate which will permit a bead to remain in contact with it throughout the oscillation.

The bead will lose contact with the plate when its acceleration is just greater than the acceleration due to gravity but has the opposite direction. For shm,

$$a = -\omega^2 x$$

therefore

$$-g = -\omega^2 x$$

Thus,

$$x = \frac{g}{\omega^2} = \frac{gT^2}{4\pi^2} = \frac{10 \times 0.3^2}{4\pi^2} = \underline{0.02 \, \text{m}}$$

Example 2.27

A simple pendulum has a period of oscillation of 2.3 s. When the pendulum is lengthened by 1 m the period becomes 3.1 s. Determine the original length of the pendulum, L, and the acceleration of free fall, g.

For a simple pendulum,

$$T = 2\pi \sqrt{\frac{L}{g}}$$

Thus,

$$T^2 g = 4\pi^2 L$$

where L is the original length of the pendulum. Thus,

$$(2.3)^2 g = 4\pi^2 L \quad \text{and} \quad (3.1)^2 g = 4\pi^2 (L + 1)$$

So $L = (2.3)^2 g / 4\pi^2$

and substituting for L gives

$$(3.1)^2 g = 4\pi^2 \left(\frac{(2.3)^2 g}{4\pi^2} + 1 \right) = (2.3)^2 g + 4\pi^2$$

Therefore

$$g = \frac{4\pi^2}{(3.1)^2 - (2.3)^2} = \underline{9.1 \, \text{m s}^{-2}}$$

Thus,

$$L = (2.3)^2 \frac{g}{4\pi^2} = \frac{(2.3)^2 \times 9.1}{4\pi^2} = \underline{1.2 \, \text{m}}$$

Example 2.28

A body oscillates with shm with a frequency of 55 Hz and an amplitude of 7 mm. What is the acceleration of the body when it is at a maximum displacement from the zero position?

For shm,

$$a = -\omega^2 x = -(2\pi f)^2 x$$

Then

$$a = -(2\pi \times 55)^2 \times 7 \times 10^{-3} = \underline{-836 \, \text{m s}^{-2}}$$

Example 2.29

Assuming no energy loss, determine the maximum PE of the body in example 2.28 if it has a mass of 1.2 kg.

Maximum speed of body occurs at zero position.

$$v_{max} = \omega A$$

Thus, maximum KE

$$E = \tfrac{1}{2}mv_{max}^2 = \tfrac{1}{2}m\omega^2 A^2$$

and so

$$E = 0.5 \times 1.2 \times 4\pi^2 \times 55^2 \times (7 \times 10^{-3}) = 3.5\,\text{J}$$

Assuming no energy loss,

$$\text{maximum PE} = \text{maximum KE} = \underline{3.5\,\text{J}}$$

Graphical treatment of damping

See Fig. 2.46.

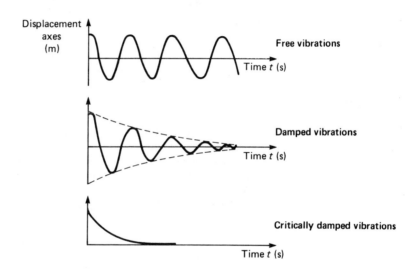

Figure 2.46

For **critical damping**, motion ceases without any oscillation. Many electrical meters are critically damped to give quick, accurate readings.

Forced vibrations

If a system is subjected to a periodic *driving force* it vibrates with the same or very nearly the same frequency as the driving force. These are **forced vibrations**. The driving force usually comes from a system in shm and is transferred from the driver to the driven by some sort of coupling.
There are three cases (Fig. 2.47).

(1)   frequency of driver < frequency of driven

The driven vibrates in phase with the driver but with a fairly small amplitude.

(2)   frequency of driver = frequency of driven

The amplitude of forced vibration is great. This is called **resonance**. The driven is out of phase with the driver by $T/4$.

(3)   frequency of driver > frequency of driven

The driven vibrates with a fairly small amplitude and is out of phase with the driver by $T/2$.

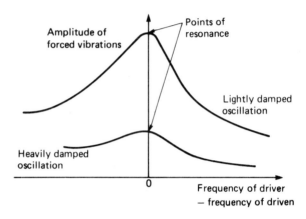

Figure 2.47

Resonance and shm

Simple harmonic motion is observed in many branches of Physics—electronics, mechanics, atomic physics and quantum physics. The reason it is called *simple* is because it is the easiest type of oscillatory motion that can be considered. It is a useful introduction to the treatment of more complicated motions.

Similarly *resonance* is found in many areas of Physics—electronics, nuclear physics, electron resonance spectroscopy and sound. Here are some examples of resonance.

(1)   A *radio circuit* is tuned by making its natural frequency of free electrical vibrations equal to that of the incoming signal.

(2)   Oscillating light waves passing through a crystal lattice force the ions in it to *vibrate* in resonance and energy is absorbed from the waves.

(3)   A loose metal plate on a car vibrates as the motor is revved up. However, when the motor reaches the natural frequency of the plate, the *amplitude* of vibration may be so great that the plate falls off the car.

(4)   Barton's pendulum consists of a number of pendulums, of varying length, attached to one long string and hanging side-by-side vertically down. Two of the pendulums are made to have the same length and when one of these is set into motion, the other will oscillate with a large amplitude. All the other pendulums have fairly small displacements. (The frequency of a pendulum is determined by its *length* and that is why resonance is shown for the two pendulums of the same length.)

(5)   Soldiers break step when marching across a bridge in case the frequency of their marching *equals* the natural frequency of the bridge and resonance occurs.

(6)   Tacoma Narrows Bridge disaster occurred as the result of violent winds setting the bridge into motion, eventually causing resonance and the collapse of the bridge.

> Resonance is the point at which the natural frequency of a body in forced vibration is equal to the frequency of the driver.

Elasticity

When a wire has a *load* acting upon it, it becomes *extended*. A graph of the force exerted on the wire versus extension reveals a region where the *extension* is directly *proportional* to the *force*. It is here that **Hooke's law** is obeyed by the wire. It suggests that when a molecule of the wire is displaced from its equilibrium position the restoring force upon it is directly proportional to the displacement. This infers that the motions of molecules in a solid are *simple harmonic* in nature.

The region of proportionality is terminated by the **limit of proportionality**, just beyond which is the **elastic limit**. Up to the latter point, the wire resumes its normal shape when the load on it is removed, i.e. it is said to be **elastic**. Both limits are close to each other and sometimes coincide.

The molecules in a wire form a lattice which contains several characteristic lattice planes. These planes have specific patterns of molecules. When the elastic limit is exceeded, the planes begin to slide over each other and, at the **yield point**, the wire starts to 'flow' and thin evenly along its length. A drastic lengthening occurs and the force on the wire reaches a maximum called the **breaking stress**. Here the wire begins to thin at a particular place more than anywhere else, and snaps.

Force—extension graph

See Fig. 2.48.

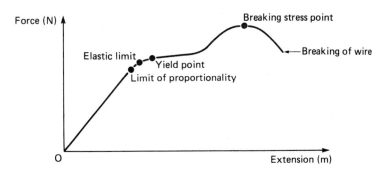

Figure 2.48

In the region after the **yield point** the wire material is said to be **plastic**.

Ductile materials are those which become plastic and lengthen considerably, e.g. copper and lead.

Brittle materials (Fig. 2.49) such as glass *fracture* just beyond the *elastic limit*. They have *no plastic region* and cannot relieve forces across surface cracks because they do not yield. Certain alloys stretch on beyond the elastic limit without becoming plastic.

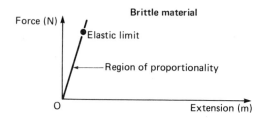

Figure 2.49

Incorrect positioning of lattice planes or the absence of an atom at a particular place determine the extent of the above properties.

In **rubber materials** (Fig. 2.50) the molecules form long chains which can be 'squashed' or coiled. They straighten when a load is placed upon the material. Under a constant load the weak molecular bonds break down in places and re-form as the chains gradually slide across each other. This enables rubber to be lengthened considerably without snapping.

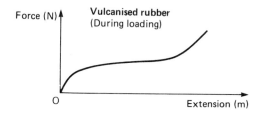

Figure 2.50

Stress and strain

Force–extension graphs are often termed stress–strain graphs. The **stress** upon a body is the force F acting per unit cross-sectional area A. The **strain** experienced by a body is the change caused in a dimension per unit dimension. For instance, in the stretching of a wire, the altered dimension is length. If the wire originally had length L and is extended by an amount e, then

$$\text{stress} = F/A \qquad \text{(units: } N\,m^{-2})$$

$$\text{strain} = e/L \qquad \text{(no unit)}$$

The ratio of stress to strain for a given material is called **Young's modulus**, which is given the symbol E.

$$E = \frac{\text{stress}}{\text{strain}} \qquad \text{(units: } N\,m^{-2})$$

Young's modulus experiment

The apparatus shown in Fig. 2.51 can be used to find Young's modulus for a *long specimen wire*. The specimen and a *reference* wire of the same length and material are suspended from chucks. This minimises the effect of temperature and the yield of the support.

Both wires are made taut and free of kinks by suitable weighting and the zero position found for the specimen wire by reading the Vernier.

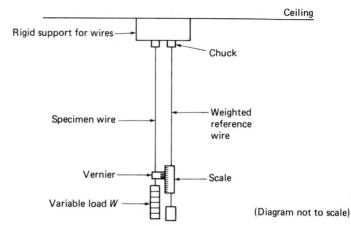

Ceiling

Rigid support for wires

Chuck

Specimen wire

Weighted reference wire

Vernier

Scale

Variable load *W*

(Diagram not to scale)

Figure 2.51

Successive Vernier readings are taken as the load on the specimen wire is increased but kept within the *elastic limit*. Readings are repeated during removal of the load, to provide mean values of extension. The length of the wire L from the point of attachment at the Vernier to the chuck is then found. A mean value for wire diameter d is obtained by making two measurements perpendicular to each other, with a micrometer, at a number of places along the wire.

If the radius of the wire is r $(= d/2)$ and the extension observed for a load W is e, then

$$\text{stress} = W/\pi r^2 \quad \text{and} \quad \text{strain} = e/L$$

Thus, Young's modulus is

$$E = \frac{WL}{e\pi r^2}$$

Also

$$e = \frac{L}{E\pi r^2} W$$

and so a graph of e versus W should be a straight line of slope $L/E\pi r^2$, from which the value of E can be determined.

Importance of Young's modulus

A material having a high value for Young's modulus experiences a smaller strain than one with a lower value under the same stress.

This is of importance in the construction and design of many engineering projects where materials are expected to withstand various stresses.

Material	$E\,(\times 10^9\,\text{N}\,\text{m}^{-2})$
Mild steel	205
Plywood	4 to 11
Brick	About 8
Concrete	About 40
Vulcanised rubber	1×10^{-3}
Lead	16

Work done when a material is stretched

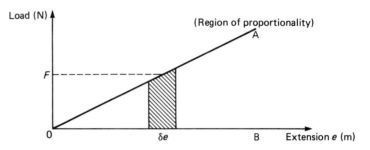

Figure 2.52

From the definition of work W,

$$W = \text{force} \times \text{distance moved in direction of force}$$

Considering only the region of proportionality for a material, when *Hooke's law is obeyed*, a very small extension δe under an average force F is achieved when a very small amount of work δW is done on the wire. Therefore

$$\delta W = F\,\delta e$$

This is the area of the small strip under the plot (see Fig. 2.52). For the total work done in stretching, all the areas of all the strips beneath the plot need to be added. That is, the area of the triangle OAB gives the total work done:

$$\boxed{W = \tfrac{1}{2}Fe}$$

An amount of energy W has now been converted to *potential energy* in the stretched material.

$$\text{stored PE in wire} = \tfrac{1}{2}Fe$$

$$\text{stress} = F/A \quad \text{and} \quad \text{strain} = e/L$$

where L and A are the original length and cross-sectional area of the wire respectively. So

$$\text{stored PE} = \tfrac{1}{2}(\text{stress} \times \text{strain})\,AL = \tfrac{1}{2}E(\text{strain})^2\,AL$$

Therefore

$$\text{stored PE} = \tfrac{1}{2}Ee^2\frac{A}{L} \quad \text{where} \quad E = \frac{\text{stress}}{\text{strain}}$$

Example 2.30

A steel wire of length $L = 6.5$ m and mass $m = 0.06$ kg is hung from the side of a tower. Determine (i) its extension and (ii) the energy stored in it, when a mass of 10 kg is attached to it. (Young's modulus for steel, $E = 2.1 \times 10^{11}$ N m^{-2}; density of steel, $\rho = 8 \times 10^3$ kg m^{-3}; $g = 9.8$ m s^{-2}.)

(i) $\text{Young's modulus } E = \dfrac{\text{stress}}{\text{strain}} = \dfrac{\text{load } W/\text{cross-sectional area } A}{\text{extension } e/\text{original length } L}$

$$\text{density} = \frac{\text{mass}}{\text{volume}} = \frac{m}{A \times L}$$

Thus,

$$A = \frac{m}{\rho \times L}$$

load $= Mg$ (M = mass attached)

Therefore

$$e = \frac{W/A}{E/L} = \frac{Mg\rho L^2}{Em} = \frac{10 \times 9.8 \times 8 \times 10^3 \times 6.5^2}{2.1 \times 10^{11} \times 0.06}$$

Thus,

$$e = \underline{2.6 \times 10^{-3}\,\text{m}}$$

(ii)   potential energy stored in wire $= \frac{1}{2}We = 0.5 \times 10 \times 9.8 \times \dfrac{2.6}{1000}$

Thus,

$$\text{energy stored} = \underline{0.13\,\text{J}}$$

Example 2.31

The limit of proportionality for a steel bar occurs for a load of 250 kN. Find the value for the Young's modulus of the steel if the strain at this limit is 0.000 03 and the cross-sectional area of the bar is 0.04 m².

$$\text{Young's modulus } E = \frac{\text{stress}}{\text{strain}} = \frac{W/A}{\text{strain}} = \frac{250 \times 10^3}{0.04 \times 0.000\,03}$$

Therefore

$$E = \underline{2.1 \times 10^{11}\,\text{N}\,\text{m}^{-2}}$$

The rod breaks at a particular point along its length when a load of 320 kN is applied. Explain what happens to the stress at this point.

$$\text{stress} = W/A = \frac{320 \times 10^3}{0.04} = 8 \times 10^6\,\text{N}\,\text{m}^{-2}$$

However, as the cross-sectional area of the rod will be smaller than usual, due to thinning, the stress will have a value greater than that calculated.

Example 2.32

A bar is suspended at each end by a steel wire of length 4 m and radius 2 mm. The bar is 5 m long and lies horizontally when a mass of 40 kg is strapped to its centre. What change in wire lengths occurs when the load is displaced to a point 1.5 m from the bar's centre? (Young's modulus for steel is $2 \times 10^{11}\,\text{N}\,\text{m}^{-2}$; $g = 9.8\,\text{m}\,\text{s}^{-2}$.)

For the load at the bar's centre, the wire tensions $T_1 = T_2$. From the *principle of moments*, when the load is moved, $4T_1 = T_2$. Thus,

$$T_1 = \frac{mg}{5} = \frac{40 \times 9.8}{5} \quad \text{and} \quad T_2 = \frac{4mg}{5} = \frac{160 \times 9.8}{5}$$

$$= 78\,\text{N} \qquad\qquad\qquad = 314\,\text{N}$$

$$\text{difference in lengths} = (T_2 L/AE) - (T_1 L/AE)$$

$$= L(T_2 - T_1)/AE$$

Thus,

$$\text{difference} = 4(314 - 78)/(\pi \times 4 \times 10^{-6} \times 2 \times 10^{11})$$

$$= \underline{4 \times 10^{-4}\,\text{m}}$$

3
Waves

Introduction

This chapter deals with the idea of a wave—a moving disturbance which transfers energy. It discusses the different types of waves and their basic properties such as wavelength, frequency and amplitude. Graphical and diagrammatic representation of waves is also illustrated.

The two major effects of reflection and refraction are then considered. The laws governing them are given and reference is made to examples of each effect from light. A discussion of total internal reflection is included here.

This is followed by a look at the ways in which waves can combine to produce interference and stationary waves. The chapter on physical optics shows how interference and diffraction (which is also mentioned here) apply to light waves. Stationary waves as applied to sound waves appear in the chapter on sound.

Polarisation is a basic property of transverse waves and thus of all electromagnetic waves. It is dealt with next and is followed by an introduction to the simple harmonic wave. The latter is a mathematical model of a wave and some of its features are given. A comparison of different wave properties concludes the chapter.

Revision targets

The student should be able to:

(1) Explain what is meant by transverse and longitudinal waves.
(2) Define wave amplitude, frequency, period and wavelength.
(3) Apply the wave equation to various numerical problems.
(4) Represent the development of a wave graphically.
(5) State the laws of reflection.
(6) Define refraction and apply Snell's law to numerical problems, possibly involving several media.
(7) Define dispersion.
(8) Discuss total internal reflection with reference to, and with examples from, light.
(9) Explain what happens when waves superpose and what is meant by interference.
(10) Define diffraction.
(11) Describe the setting up and properties of stationary waves.
(12) Explain the term plane polarisation.
(13) Give the equation for a simple harmonic wave in terms of the phase angle and in terms of the wavenumber.

(14) Apply the equation for a simple harmonic wave to simple numerical problems.

(15) Compare and give examples of the properties of major types of waves, including light and sound.

Progressive wave

The term 'progressive' applied to a wave means **travelling**. Such a wave can be thought of as a disturbance that moves away from a source to surrounding places resulting in the *transfer of energy*. There are two main types of wave: transverse progressive and longitudinal progressive waves.

Transverse progressive

In this type of wave motion the vibrations occur *at right angles* to the direction of propagation of the wave. (To propagate means to spread.)

Longitudinal progressive

These are waves in which the vibrations actually occur *in the direction* of propagation of the wave itself.

The diagrams in Fig. 3.1 show the passage of a wave through a medium consisting of a chain of particles.

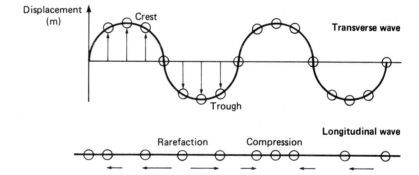

Figure 3.1

Wave properties

A number of terms which apply to periodic motion, like that of a pendulum for example, also apply to wave motion. They have the same symbols and units.

Wave amplitude is the maximum value of an alternating quantity from equilibrium. This quantity might be for example the displacement of water droplets in a surface water wave.

Wave frequency is the number of cycles per second.

Wave period is the time taken for one complete cycle.

Wavelength—the wavelength λ of a wave is the distance between two neighbouring points on the wave which are exactly in phase with each other. (In phase means 'in step'.) See Fig. 3.2.

After a time t the wave moves a certain distance. After a time T the wave moves along by a distance of one wavelength. So the *velocity* of the

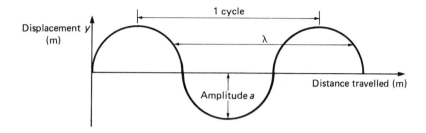

Figure 3.2

wave v is given by

$$v = \lambda / T$$

and so

$$\boxed{v = f\lambda}$$ (f = wave frequency)

Mechanical waves

These are produced by a disturbance in a medium and are transmitted by particles of that medium oscillating to and fro. Examples are surface water waves, waves on a stretched string, and sound waves.

Electromagnetic waves

These are in the form of varying electric and magnetic fields. They can travel through a medium or a *vacuum*. Examples are light waves, radio waves, and X-rays.

Graphical representation of waves

There are two ways of representing waves graphically. In one a snapshot is considered, to show what the wave is doing at various distances from its source at a *particular instant*. In the other a particular point is chosen, and what happens to the wave as time goes on at that point is considered. For example, see Fig. 3.3.

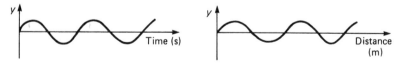

Figure 3.3

The quantity y *can be* displacement, as for example in a mechanical wave, or the magnitude of a field, as in the electromagnetic waves.

Wavefronts and rays

A wavefront is a line or surface on which the disturbance has the same *phase* at all points. A line at right angles to a wavefront which shows its direction of travel is a ray. See Fig. 3.4.

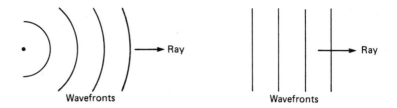

Figure 3.4

Huygen's construction

Each point on a wavefront can be regarded as a source of *secondary wavelets*, which spread out with the wave velocity. The new wavefront is the surface which touches all the wavelets. Arcs of wavelets are only drawn in the *forward* direction. See Fig. 3.5.

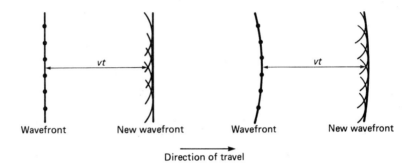

Figure 3.5

Laws of reflection

First law—the angle of incidence is always equal to the angle of reflection.
Second law—the normal to the reflecting surface, the incident ray and the reflected ray all lie in the same plane.

Angles of incidence, reflection, etc., are always drawn with respect to the line perpendicular to the surface concerned, called the *normal* (Fig. 3.6).

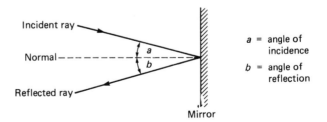

Figure 3.6

Refraction

This is the deviation of a wave when it goes from one medium to another in which it has a different *speed* (Fig. 3.7).

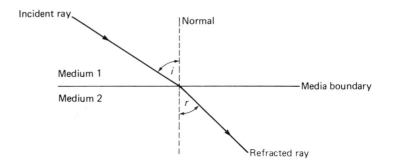

Figure 3.7

Snell's law of refraction

The ratio of the sine of the angle of incidence of a wave to the sine of the angle of its refraction is a *constant* called the **refractive index**, for a given pair of media.

Thus, if i = angle of incidence and r = angle of refraction, then Snell's law can be written as

$$\frac{\sin i}{\sin r} = {}_1n_2$$

The constant ${}_1n_2$ is the refractive index for a wave travelling *from* medium 1 *to* medium 2. It can be shown that the refractive index is related to the speeds of the wave v_1 and v_2 in media 1 and 2 respectively as follows:

$$_1n_2 = \frac{v_1}{v_2}$$

The **absolute refractive index** n_m of a medium m is defined as the ratio of the speed of the wave in a vacuum to its speed in the medium m. Thus, the refractive index may also be expressed as

$$_1n_2 = \frac{n_2}{n_1}$$

(The absolute refractive index for air is one.)

Wave direction and refraction

From Snell's law, $_1n_2 = v_2/v_1$ for a wave moving from medium 1 to medium 2.

Similarly, $_2n_1 = v_1/v_2$ for a wave moving from medium 2 to medium 1. Finally, therefore,

$$_1n_2 = \frac{1}{_2n_1}$$

The refractive index indicates how strongly a medium will deviate waves. The larger its value, the greater is the deviation.

From Snell's law, $\sin i = {}_1n_2 \sin r$, and this is of the form $y = mx + c$. Thus, by making measurements of the angle of incidence i and the angle of refraction r, a graph of $\sin i$ versus $\sin r$ should give a straight line passing through the origin whose *gradient* is the refractive index.

Laws of refraction

First law—Snell's law.
Second law—the incident ray, the refracted ray and the normal to the refracting surface all lie in the same plane.

Refraction through several media

Suppose waves pass from medium 1 to medium 2, then into medium 3 and back into medium 1. This is shown in Fig. 3.8.

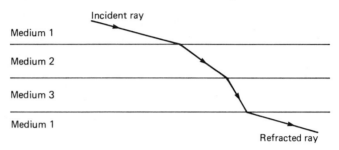

Figure 3.8

Then
$$_1n_2 = n_2/n_1 \quad \text{and} \quad _3n_1 = n_1/n_3$$

Therefore
$$_3n_2 = \frac{n_2}{n_1} \times \frac{n_1}{n_3}$$

Thus,
$$_3n_2 = {}_1n_2 \times {}_3n_1$$

Dispersion

The speed of a wave in a medium depends on its *wavelength*. For most transparent substances the refractive index increases as the wavelength decreases. Thus, waves of short wavelength are usually refracted less than those of longer wavelength. If this happens in a medium it is said to be *dispersive*.

Multiple images in mirrors

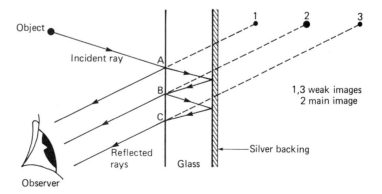

Figure 3.9

A thick glass mirror will produce a number of images from a single object. In Fig. 3.9 light is refracted at A but a small amount is *reflected* which produces a *weak* image. On reflection at B the ray is refracted back into the air forming the *main* and brightest image. Further reflection occurs at C producing yet another *weak* image.

The total effect is seen as the *blurring* of the main image.

Total internal reflection

This is an effect which occurs when light goes from one medium to another which is optically less dense, and can be usefully applied in overcoming the problems of thick mirrors. A semicircular glass block is used to demonstrate total internal reflection in the laboratory.

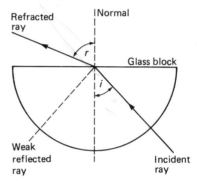

Figure 3.10

As the angle *i* increases from zero, the angle of refraction *r* also increases (Fig. 3.10). Apart from the refracted ray there is also a weak reflected ray. When angle *i* reaches a value called the *critical angle c*, the angle of refraction is 90° (Fig. 3.11). If the angle of incidence is made *larger* than *c*, all the light will be reflected inside the glass block, i.e. **total internal reflection** occurs.

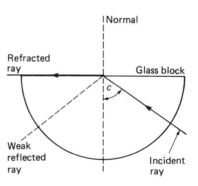

Figure 3.11

From Snell's law, at the critical angle of incidence

$$_g n_a = \frac{\sin i}{\sin r} = \frac{\sin c}{\sin 90°}$$

But $\sin 90° = 1$, therefore

$$_g n_a = \sin c$$

Also,

$$_a n_g = \frac{1}{_g n_a}$$

So

$$_a n_g = \frac{1}{\sin c}$$

Totally reflecting prisms

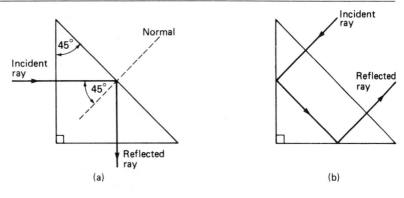

Figure 3.12   (a)   (b)

The *critical angle* of glass is about 42°. Thus, a ray of light striking the longest face of a 45° prism as shown in Fig. 3.12(a), with an angle of incidence of 45°, will be totally internally reflected.

Light can also be reflected through 180° by a prism as shown in Fig. 3.12(b), and this is used in the design of binoculars.

As prisms cause *no blurring* of the image, which *does* occur in thick glass mirrors, they are used as reflectors in precision optical instruments, for example *periscopes* (Fig. 3.13).

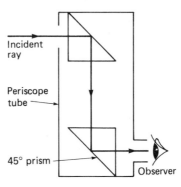

Figure 3.13

Light pipes Light can be trapped inside a fine glass fibre by total internal reflection and then sent along it (Fig. 3.14). This has applications in medicine and engineering where internal parts can be illuminated for inspection. It also has applications for telephony, where thousands of calls can be passed down a thin cable of optical fibres.

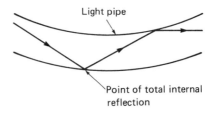

Light pipe

Point of total internal
reflection

Figure 3.14

Mirages   Blue light from the sky is gradually refracted by warmer layers of air, on its way down. When it is incident on a less dense layer of air with an angle *greater* than the critical angle, total internal reflection occurs. Such a layer is usually found close to the ground which has been heated by the Sun. An observer thus sees blue light apparently coming off the ground and imagines water to be there. (Examples of heated surface include desert sand and tarmac roads.)

Total internal reflection of radio waves

A layer of atmosphere containing a large number of electrons called the Appleton layer refracts radio waves coming from the Earth's surface until the critical angle is exceeded. Total internal reflection then occurs and this enables radio and TV signals to be received despite the curvature of the planet.

Example 3.1

Light passes into parallel sided layers of liquid jelly and perspex. If the refractive indices for liquid jelly and perspex are 0.76 and 0.59 respectively, determine the following for an angle of incidence of $6°$ at the air/jelly interface: (i) the angle of refraction in the jelly, (ii) the angle of incidence at the jelly/perspex interface, and (iii) the angle of refraction in the perspex layer.

(The subscripts a, j and p are used for air, jelly and perspex respectively.)

(i)   From Snell's law,

$$_a n_j = \frac{\sin 6°}{\sin r_j} = 0.76$$

Thus, $\sin r_j = \dfrac{\sin 6°}{0.76}$

and so

$$r_j = \underline{7.9°}$$

(ii)   Angles r_j and i_j are alternate angles and so $i_j = \underline{7.9°}$ too.

(iii)   For three media 1, 2 and 3 the following relationship applies:

$$_3 n_2 = {}_1 n_2 \times {}_3 n_1$$

Thus, $_j n_p = {}_a n_p \times {}_j n_a = {}_a n_p \times \dfrac{1}{{}_a n_j}$

Then, $_j n_p = 0.59 \times \dfrac{1}{0.76} = 0.78$

Therefore
$$_j n_p = 0.78 = \frac{\sin 7.9°}{\sin r_p}$$
and so
$$r_p = \underline{10.2°}$$

Example 3.2

If the refractive indices of materials X and Y are 1.28 and 1.41 respectively, explain under what conditions total internal reflection will occur.

Since the refractive index for Y is greater than that for X, then only rays of light from Y entering X will have a chance of suffering total internal reflection. Now.

$$_Y n_X = \,_{air} n_X \times \,_Y n_{air} = \,_{air} n_X \times \frac{1}{_{air} n_Y}$$

Therefore
$$\sin c = \,_Y n_X = 1.28 \times \frac{1}{1.41} = \underline{0.91}$$

Thus, light incident in Y at the Y/X interface with an angle greater than $c = 65.6°$ will be totally internally reflected.

Example 3.3

The refractive index for a wave passing from medium A to medium B is 0.67. If the speed of the wave in medium A is 3.45×10^3 m s^{-1}, what is its speed in medium B?

Refraction is the deviation of a wave as it passes from one medium to another due to a *change in its speed*.

Now from Snell's law,

$$_A n_B = \frac{v_A}{v_B}$$
and so
$$v_B = \frac{v_A}{_A n_B}$$

Thus, $v_B = 3.45 \times 10^3/0.67 = \underline{5.15 \times 10^3 \text{ m s}^{-1}}$

Principle of superposition of waves

When two waves travel in a medium, the combined effect at any one point can be found from this principle: the resultant displacement at any point is the sum of the separate displacements due to the waves.

Interference

When waves from *coherent* sources cross, superposition occurs giving rise to reinforcement or cancellation. The result is called an **interference pattern** or **interference fringes**.

Coherent sources have a *constant phase difference* and therefore the same frequency. For **incoherent sources**, reinforcement and cancellation will alternate at any one place to give an average effect. An analogy might be a crowd of people all saying the same sentence at the same time. Every word can be heard clearly. These are coherent sources. If each person says the sentence at random, only garble is heard. These are then incoherent sources. By using Huygens' construction, wavefronts produced by two point sources A and B can be shown to exhibit interference (Fig. 3.15).

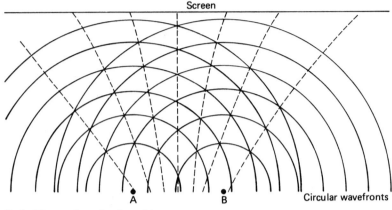

Figure 3.15

Dashed line marks antinode positions

Where the wavefronts cross they are in phase, and their individual amplitudes add up to give reinforcement (Fig. 3.16). In between these points cancellation occurs as the wavefronts are 180° out of phase

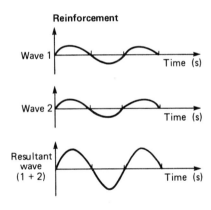

Figure 3.16

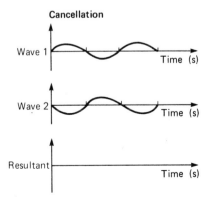

Figure 3.17

(Fig. 3.17). A region where cancellation takes place is called a **node** while a region where reinforcement arises is called an **antinode**.

Interference can be demonstrated by placing a 3 cm wave source behind two narrow slits. A detector on the other side of the slits picks up maxima and minima as it is moved across them.

Diffraction

This is the spreading of waves as they pass through an opening or travel past an obstacle. It can be demonstrated with 3 cm waves. Two vertically placed metal plates (A and B) form a slit (Fig. 3.18). If the detector is placed on the other side of the slit, away from the transmitter, just behind plate A, then as B is moved closer to A, the signal gets stronger. It is strongest when the aperture width is about 3 cm—the wavelength of the waves used. This effect is due to destructive interference being gradually removed from point sources on the wavefront going through the aperture.

The wavelength of light is 6×10^{-7} m and so diffraction around ordinary objects is small. Thus, it is not possible to 'see' around corners.

The wavelength of sound at ordinary temperatures is about 0.5 m, and so it is possible to 'hear' around corners.

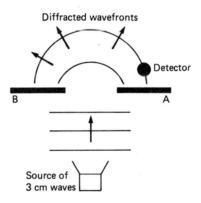

Figure 3.18

Stationary waves

These are formed by two waves of equal amplitude and frequency travelling in opposite directions (see Fig. 3.19). They stay in the same place in the medium. Their properties include the following.

(1) *Permanent* nodes.
(2) In one 'loop' all the particles oscillate with *different amplitudes* but are *in phase* with each other. (In a plane progressive wave the amplitude stays the same at all points throughout the medium, but neighbouring particles are not in phase with each other.)
(3) The oscillations in one 'loop' are in *antiphase*, i.e. 180° out of phase, with those in the following 'loop'.
(4) The frequency of particle vibration, and the wavelength of the stationary wave, is the same as in the progressive waves which were used to make the stationary wave.

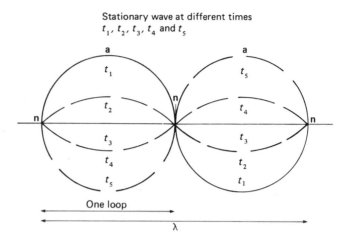

Stationary wave at different times
t_1, t_2, t_3, t_4 and t_5

Figure 3.19

Stationary waves can be demonstrated with 3 cm waves by using a vertical metal plate to reflect waves back to a source. A microphone moved along the line between the plate and the source will give a series of maxima and minima. Boundaries are needed for stationary waves as the progressive waves reflected from them interfere and the resulting waves have to 'fit' into the system. As energy is not transmitted, large amounts of energy can be trapped in these waves. If by the time a reflected wave returns to the source, the source has produced a whole number of more waves, resonance occurs as the waves meet in phase and the *amplitude increases*. This effect therefore must be minimised in devices such as turbines and propellers.

Polarisation

This only occurs with *transverse waves*, e.g. electromagnetic waves. The displacements of transverse, surface water waves (e.g. in a ripple tank) are always perpendicular to the water surface, i.e. up or down. See Fig. 3.20.

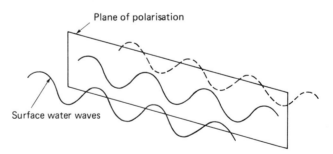

Plane of polarisation

Surface water waves

Figure 3.20

The waves are said to be **plane polarised**. However, the oscillations in electromagnetic waves can be in any plane normal to the direction of propagation (Fig. 3.21). They are **unpolarised**.

Electromagnetic waves are made up of two oscillating fields perpendicular to each other (Fig. 3.22). These are electric and magnetic fields.

The magnitudes of the fields are represented by E and B. If the electromagnetic wave becomes polarised, the electric field always

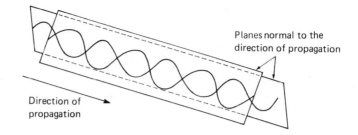

Figure 3.21

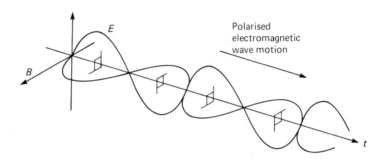

Figure 3.22

oscillates in one plane (the plane of polarisation) and the magnetic field always oscillates in a plane perpendicular to it.

A vertical aerial produces a vertically plane polarised electromagnetic wave with the electric field vertical. This is the way in which 3 cm waves are produced, and if a series of thin metal bars is placed in front of the source, polarisation can be demonstrated. If the bars are vertical, the vertical electric field excites electrons in the bars and no signal reaches the detector. When the bars are horizontal, only a few electrons are excited and most of the signal is obtained.

Polaroid lenses polarise sunlight. If two are placed together and rotated, a position will be reached where plane polarised light from one is totally eliminated by the polarising effect of the other.

Reflection and phase change

A *phase change* of 180° occurs when a transverse wave is reflected at a denser medium. There must be a node at the boundary and only waves travelling in opposite directions which are 180° out of phase combine to give a *node*. 3 cm waves can be used to demonstrate this by pointing a source of the waves at a vertical metal plate. If a microphone is moved up and down as shown in Fig. 3.23, maxima and minima will be detected. Waves travelling direct to the microphone interfere with the waves which are reflected. At P a maximum should occur but, because of the phase change, a minimum arises.

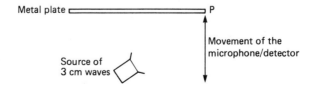

Figure 3.23

Applications   Radar and sonar are used to detect the positions of aircraft, submarines etc. A wave is transmitted and its reflection noted. The total wave travel time $t = 2 \times$ time taken to reach object. So the distance to the object $= vt/2$ where v is the speed of the wave.

Simple harmonic waves

When considering simple harmonic *motion* it was found that a harmonically varying quantity y (e.g. displacement from a fixed point) can be written as

$$y = a \sin \omega t$$

where a is the amplitude.

This can now be used to obtain expressions for a *progressive* or *travelling wave* (Fig. 3.24).

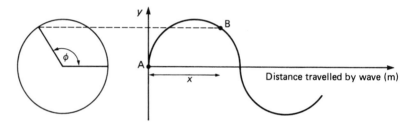

Figure 3.24

A wave travels from a particle A to a particle B with speed v. B will receive the disturbance a time x/v later than A did.

So B only oscillates for a time $t - x/v$ and so the displacement y can now be written as

$$y = a \sin \omega(t - x/v)$$

If the **phase angle** (sometimes called the **epoch angle**) goes from zero to ϕ, the wave travels a distance x. If the phase angle goes from zero to 360° (2π radians), the wave travels a distance of one wavelength λ. Therefore

$$x = \lambda\phi/2\pi$$

Note: The term above in brackets is *also* often referred to as the phase angle; fortunately, however, '*the phase of the wave*' is another name for it!

Equations for simple harmonic waves

Recalling that the angular velocity $\omega = 2\pi v/\lambda$, then the basic equation for a simple harmonic wave:

$$y = a \sin \omega(t - x/v)$$

can also be written in the following ways:

$$y = a \sin \left(\omega t - \frac{2\pi x}{\lambda} \right)$$

or as

$$y = a \sin (\omega t - \phi)$$

The term $2\pi/\lambda$ is often written as k which is called the **wavenumber**. Therefore

$$\boxed{y = a \sin(\omega t - kx)}$$

For a wave moving from particle B back to particle A, the minus sign inside the brackets becomes a plus sign, if a similar analysis is performed.

$$y = a \sin(\omega t + kx)$$

By examining the equations above it is possible to see that there are *two ways* of graphically representing a wave.

If t, the time, is set to a particular value (i.e. made a constant), then a 'snapshot' of the wave in time is being considered.

If the distance travelled, x, is made a constant, the variation of y with respect to time at a certain place is being considered.

Equation for a stationary wave

The displacement y of a stationary wave can be obtained by applying the definition for such a wave. It is a wave formed by two waves of equal amplitude and frequency travelling in opposite directions.

$$y = a \sin(\omega t - kx) + a \sin(\omega t + kx) = 2a \cos kx \cdot \sin \omega t$$

The amplitude of the stationary wave at any point along it is given by $2a \cos kx$. It can be seen that this is zero (i.e. a *node* occurs) every *half* a wavelength along the wave.

Wave effects

A few examples of the applications of wave properties are tabulated below.

Wave property	Light	Other electromagnetic wave	Sound
Reflection	Car mirror	HF radio, radar	Sonar
Refraction	Prism	IR—using a magnifying glass to burn skin	Sound travels further at night
Interference	Colours in soap bubbles	Radio-interference	The production of 'beats'
Diffraction	Images of street lamp through eye-lashes	X-ray analysis of crystalline materials	Ability to hear round large obstacles
Polarisation	Polarimeter	Radio, TV antenna	Does not occur
Energy transfer	Solar power	Gamma rays—treatment of cancer	Using ultrasonics to dislodge dirt

Example 3.4

A guitar string is stretched between two bridges a distance L apart. If it is caused to vibrate in the lowest possible frequency which will produce an antinode a distance $L/6$ from one bridge, find the next, higher frequency which will produce an antinode in the same position. (The lowest frequency is 120 Hz.)

For the lowest frequency of stationary wave to produce an antinode a distance $L/6$ from one bridge, each $L/6$ distance along the string must contain a quarter wavelength (Fig. 3.25).

Figure 3.25

For the next, higher frequency, each $L/6$ distance along the string will need to contain $3/4$ of a wavelength. Thus, the distance between the bridges will be taken up by 4.5 wavelengths. The lowest frequency mode only contains 1.5 wavelengths in the distance L. Thus, the frequency has trebled, i.e.

$$\text{next, higher frequency} = 3 \times 120 = \underline{360 \text{ Hz}}$$

Example 3.5

A plane progressive wave travels through a medium. If the amplitude of the wave is 3.4×10^{-5} m and it has angular frequency 5.7×10^2 rad s^{-1} and a wavenumber $k = 20$ rad m^{-1}, find (i) the wave frequency and speed and (ii) the greatest value of speed for a molecule in the medium as the wave passes through.

(i)   Angular frequency $\omega = 2\pi f$. Thus, $f = \omega/2\pi$. Therefore

$$f = \frac{5.7 \times 10^2}{2\pi} = \underline{91 \text{ Hz}}$$

wavenumber $k = 2\pi/\lambda$

Thus, $\lambda = 2\pi/k$ and   speed $v = f\lambda$

Therefore

$$v = f \times \frac{2\pi}{k} = 91 \times \frac{2\pi}{20} = \underline{29 \text{ m s}^{-1}}$$

(ii)   Assuming that the wave passing is a simple harmonic one, then each molecule is set into simple harmonic motion. The greatest speed of a body in shm is given by

$$v_{\text{max}} = \omega A = 5.7 \times 10^2 \times 3.4 \times 10^{-5} = \underline{0.02 \text{ m s}^{-1}}$$

Example 3.6

The displacement of molecules in a medium produced by the passage of a simple harmonic wave through it is given by

$$y = 0.47 \sin (23t - 16x)$$

Determine (i) the wave amplitude, (ii) the wavelength, (iii) the wave speed and (iv) the phase difference between molecules 0.5 m away from each other.

(i)   amplitude $= \underline{0.47 \text{ m}}$

(ii)   Since $y = A \sin(\omega t - kx)$, then

$$\lambda = 2\pi/k = 2\pi/16 = \underline{0.39 \text{ m}}$$

(iii) $v = f\lambda = \dfrac{\omega\lambda}{2\pi} = \dfrac{23 \times 0.39}{2\pi} = \underline{1.4 \text{ m s}^{-1}}$

(iv)   Molecules which are 2π out of phase are one wavelength apart. Thus, molecules 0.5 m apart have a phase difference of

$$0.5 \times 2\pi/0.39 = 1.28 \times 2\pi \quad \text{or} \quad \underline{2.56\pi \text{ rad} = 461°}$$

4
Sound

Introduction

This chapter concerns itself with the properties of sound waves and expands some of the work covered in the previous chapter.

It begins with a mention of the range of sonic frequencies and the effects, such as pressure variation, which occur when a sound wave passes through a medium. The ways in which reflection, refraction, interference and diffraction arise in sound are then looked at in turn.

This section is followed by a consideration of beats (two notes of slightly different frequencies sounded together) and a look at the properties of musical notes. Stationary waves in pipes and waves on strings are described next. An application of the latter is the sonometer where the waves on the wire produce sound waves in the surrounding air.

The next part of the chapter looks at the speed of sound in various types of media and its experimental determination. Two methods are described and both are applications of stationary waves.

Finally, the chapter is concluded with a discussion of the theory behind the Doppler effect in sound where an apparent frequency is produced as a result of the relative motion between an observer and the source of the sound.

Revision targets

The student should be able to:

(1)   Describe the main properties of sound.
(2)   Explain pressure and displacement variation in a medium due to the passage of sound waves.
(3)   Discuss the superposition of sound waves with reference to beat formation.
(4)   Apply the equation for beat frequency to numerical problems.
(5)   Define the characteristics of a musical note.
(6)   Investigate problems dealing with the vibrations of gas columns in open and closed pipes.
(7)   Explain quantitatively the vibration of a string fixed at both ends and the factors which determine its fundamental frequency.
(8)   Give accounts of the experimental determination of the speed of sound in free air and in gas columns.
(9)   Explain what is meant by the Doppler effect and determine the apparent frequency of sound resulting from the motion of the source and/or the observer.

Nature of sound

Sound waves cannot be polarised and do not travel in a vacuum. They are **longitudinal** waves, and are produced by a vibrating system which imposes an oscillatory to and fro motion, along the line of wave travel, upon the motion of the particles of the transmitting medium. The speed of sound in air at room temperature is about 330 m s^{-1}.

Sonic spectrum

Region	Example	Frequencies (Hz)
Infrasonic	Earthquake	0–20
Audible	Sound	20–20 000
Ultrasonic	Dog-whistle	20 000 +

Using the **wave equation** $v = f\lambda$, the audible wavelength range is found to be 17 mm–17 m. The sonic spectrum can be investigated using a loudspeaker fed by a signal generator. A microphone is used to display the signals on an oscilloscope. At high frequencies, although nothing is heard, there still is a trace on the oscilloscope. (Ultrasonics are used in cleaning metal surfaces, e.g. watchmaker's bath, and bats' navigation systems.)

Sound is transmitted through air and other gases by a series of **compressions** and **rarefactions**. Air molecules at the centres of compressions and rarefactions have zero displacement. The graph in Fig. 4.1 indicates molecular displacements to the right and the left. The speed of the molecules (dashed curve) is a maximum at the centres of compressions and rarefactions. (The amplitudes of displacement are very small—about 0.01 mm.)

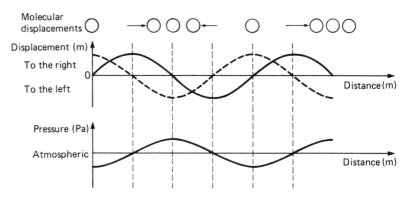

Figure 4.1

The bunching of molecules in some places and not in others causes a variation in *pressure* and *density*. The variation of pressure away from atmospheric is also very small—about 30 Pa.

Sound is best *reflected*, like other waves, by surfaces which match the wavelength of the sound, in terms of flatness, and thus rough surfaces reflect it well, giving rise to echoes. Surfaces in auditoria need to be designed carefully because persistence of sound, i.e. reverberation, can spoil the effect of a performance.

Sound waves are also *refracted* and obey Snell's law. Noise travels further at night because layers of air, heated by day, have risen. Refraction by them is greater than by cooler layers.

Interference in sound

Sound waves interfere to give *fringes* if *coherent* sources are available, and this can be demonstrated by using two loudspeakers connected in *parallel* with a signal generator. A microphone connected to a pre-amplifier and an oscilloscope will indicate maxima and minima when it is moved from side to side between the speakers.

Diffraction

An example of the diffraction of sound waves is the ability to hear round corners.

Beats

These occur when two notes of *slightly different* frequencies, but similar amplitudes, are sounded together. The loudness increases and decreases *periodically*. This can be demonstrated using two signal generators, each feeding a loudspeaker, or by using two tuning forks of the same frequency, but one with a piece of plasticine on it.

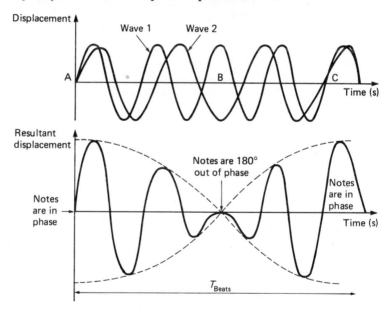

Figure 4.2

In this case the sources are *not quite coherent* and so at a particular point there is sometimes reinforcement and sometimes cancellation.

At point A in Fig. 4.2 both notes are in phase and the *resulting amplitude* is the *sum* of the amplitudes of the notes. But, as the notes have slightly different frequencies, they begin to get out of phase. At B they are half a cycle out of phase and cancellation occurs. At C they have a phase difference of exactly one cycle and are again in phase.

Beat frequency

Suppose that two slightly different wave frequencies are f_1 and f_2 where $f_1 > f_2$ and the *period of the beats* is T.

Therefore when the two sources of f_1 and f_2 are *in phase* there will be a beat, and after a time T there will be *another* when they are *in phase again*. One of the waves has gained *one* cycle in the time T. So

$$f_1 T - f_2 T = 1$$
$$f_1 - f_2 = 1/T$$

Therefore the frequency of the beats is

$$\boxed{f = 1/T = f_1 - f_2}$$

Noise

This is an audible sound which is not harmonious to the ear. Its waveform is *non-periodic*.

Musical note

This is an audible sound which is pleasant to hear and its waveform is periodic. A note has three characteristics: loudness, pitch, and quality or timbre.

Loudness

This depends on the listener and is determined by the **intensity** of the sound and the **sensitivity** of the ear.

The *intensity* is proportional to the *energy* of the sound wave and to the mass of air vibrating. Motions of air layers can be expressed in terms of simple harmonic motion, even though they are not. For shm,

$$y = a \sin \omega t$$
$$v = \frac{dy}{dt} = a\omega \cos \omega t$$
$$\text{kinetic energy} = \tfrac{1}{2} mv^2 = \tfrac{1}{2} a^2 m \omega^2 \cos^2 \omega t$$

Therefore the kinetic energy is proportional to the square of the amplitude and also to the mass. Since intensity is proportional to the energy, loudness is also proportional to the *square of the amplitude* and to the *mass of air vibrating*.

Pitch

This is also a sensation and depends on the frequency of the air which is the same as that of the source of the sound.

high frequency . . . high pitch
low frequency . . . low pitch

Pitch in sound is similar to *colour in light*. An **octave** is the interval between two notes, the *fundamental components* of which have a frequency ratio of 2:1.

Quality or timbre

Most sounds are made up of vibrations of more than one frequency. A note consists of a **fundamental** of greatest intensity and lowest frequency and several **overtones** of lesser intensity. Their frequencies are simple *multiples* of that of the fundamental.

Instruments produce notes which have different overtones and so the quality of the note is different. **Fourier analysis** is a mathematical way of taking a complex waveform and breaking it up into its fundamental and overtones. These are sinusoidal in form (Fig. 4.3).

Fundamental
(1st *harmonic*) *f* Hz

1st overtone
(2nd *harmonic*) 2*f* Hz

2nd overtone
(3rd *harmonic*) 3*f* Hz

Figure 4.3

Vibrating air columns

Here, progressive sound waves travel from a source (tuning fork, reed, etc.), to the end of a pipe where they are reflected and interfere with the incident waves. The wavelength of some waves will be such that the length of the pipe produces **resonance**, and the amplitude of these waves is large. As a result, a *stationary longitudinal wave* is formed.

Closed pipes

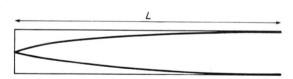

Figure 4.4

At the closed end there is a **node** and at the open end an **antinode**; see Fig. 4.4. So for the *fundamental*

$$L = \lambda/4$$

Therefore

$$f = \frac{v}{4L} \quad \text{(since } f = v/\lambda)$$

where v is the speed of the sound in the air.

In order to meet the two *boundary conditions* above, only whole loops (i.e. $\lambda/2$) can be added (Fig. 4.5).

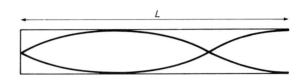

Figure 4.5

Now

$$L = \frac{\lambda}{4} + \frac{\lambda}{2} = \frac{3\lambda}{4}$$

Therefore

$$f = \frac{3v}{4L}$$

Adding another loop:

$$L = \frac{3\lambda}{4} + \frac{\lambda}{2} = \frac{5\lambda}{4}$$

Therefore

$$f = \frac{5v}{4L} \quad \text{and so on}$$

So *closed* pipes only give *odd harmonics*, i.e. fundamental, 3rd, 5th, ...

Open pipes

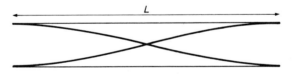

Figure 4.6

Here there are only **antinodes** at the ends; see Fig. 4.6. For the **fundamental** $L = \lambda/2$ and $f = v/2L$. Again to meet the *boundary conditions* only $\lambda/2$ can be added (Fig. 4.7).

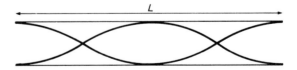

Figure 4.7

Then

$$L = \frac{\lambda}{2} + \frac{\lambda}{2} = \lambda$$

So

$$f = \frac{v}{L}$$

Adding another loop:

$$L = \lambda + \frac{\lambda}{2} = \frac{3\lambda}{2}$$

So

$$f = \frac{3v}{2L} \quad \text{and so on}$$

So *open ended* pipes give *all the harmonics*.

Further points on vibrations in pipes

(1) The note from an open pipe is of a higher quality than that from a closed one, due to the presence of the extra overtones.
(2) Actual vibration of air in a pipe is the superposition of the many modes possible, i.e. several modes can occur simultaneously.
(3) Due to vibration just at the edge of a pipe opening, the length of the pipe can be corrected to $L + 0.6r$ where r is the pipe radius.

Vibrating strings

When a stretched string or wire is bowed or plucked, progressive transverse waves travel to the fixed ends where they are reflected and then are superimposed onto the incident waves. A stationary wave is then formed when their wavelengths 'fit' the string length and resonance in the system occurs. If the ends are fixed then there must always be nodes there (Fig. 4.8).

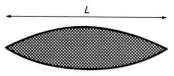

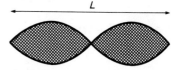

Figure 4.8

String plucked at centre gives the fundamental, and as $L = \lambda/2$ then $f = v/2L$

String plucked at $L/4$ from one end gives the 1st overtone, and as $L = \lambda$ then $f = v/L$

The *quality* of the note on a string depends on the overtones and where the string is plucked. The *speed* of a *transverse* wave on a string can be given in terms of the tension in the string T and the mass per unit length of the string μ as follows:

$$v = \sqrt{T/\mu}$$

and since

$$v = f\lambda$$

$$f = \frac{1}{\lambda}\sqrt{\frac{T}{\mu}}$$

Thus the **frequency** is inversely proportional to the wavelength (and hence to the length of the string as well), directly proportional to the square root of the tension, and inversely proportional to the square root of the mass per unit length. This is provided that in each case the other variables remain constant. (The mass per unit length is an indication of thickness for any particular type of string.)

These relationships can be confirmed by use of a **sonometer**, which is a wire fixed at one end, made taut by suspending a weight from the other end, and placed over wooden bridges on a hollow wooden box (Fig. 4.9).

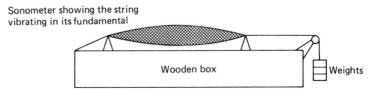

Sonometer showing the string vibrating in its fundamental

Wooden box   Weights

Figure 4.9

A demonstration of waves on a string can be provided by applying a.c. to the sonometer wire which produces an alternating magnetic field. A *perpendicular* magnetic field provided by a permanent magnet will force the sonometer string to vibrate. As the length of the vibrating string is increased, there comes a point when it vibrates as one loop, i.e. in its fundamental. At this point, the system is **resonating**. The natural frequency of the wire equals the frequency of the applied a.c. and thus the latter can be determined. The experiment can then be repeated using different tensions.

Frequency determination

Frequency can be measured by using a **stroboscope** which is basically a light which flashes on and off with *variable* frequency. If its frequency matches that of the vibrating system which it is being used to illuminate, then that system appears to be still.

Speed of sound

A *mechanical* wave is transmitted by the vibration of particles within the medium through which it is passing, and such a system will only vibrate if it has *mass and elasticity*. If E is Young's modulus of elasticity for a solid and ρ is its density then the speed of sound through that solid is given by

$$v = \sqrt{E/\rho}$$

If P is the pressure of a gas, ρ its density and γ the ratio of the specific heat capacities of the gas then the speed of sound in a gas is given by the following:

$$v = \sqrt{\gamma P/\rho}$$

For *one mole* of *ideal gas*

$$P = RT/V$$

and from the definition of density, where M is the mass of the ideal gas,

$$v = \sqrt{\gamma RT/M}$$

and so

$$v \propto \sqrt{T}$$

Thus the speed of sound in a gas is *independent* of the gas *pressure*. (Increase in pressure causes a corresponding increase in density.)

Experimental determination of the speed of sound

In free air by stationary waves

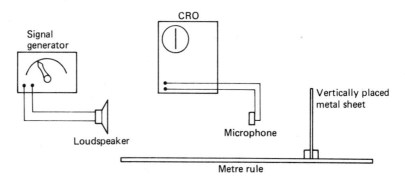

Figure 4.10

A sound of known frequency is produced from a loudspeaker attached to a signal generator (Fig. 4.10). It forms a stationary wave by reflection off a vertically placed metal sheet. If the sheet is moved away from the loudspeaker, then the trace produced by a microphone connected to the Y plates of an oscilloscope (zero timebase) will vary in amplitude from a maximum to zero then a maximum and so on. The maxima correspond to antinodes and so if the average distance between two maxima is obtained this will give half the wavelength. Thus by applying the wave equation, the speed can be determined.

Alternatively, if the sheet is removed and the loudspeaker connected to the X plates of the CRO then as the microphone is moved away from the loudspeaker, the *phase difference* between the two inputs to the CRO will vary producing different *Lissajous figures*.

Resonance tube experiment

Here a *closed pipe* is made by placing a glass tube into a large cylinder of water (Figs 4.11, 4.12). (Versions with tubes connected to reservoirs of water—whose height can be varied, thus altering the water level in the glass tube—are also used.)

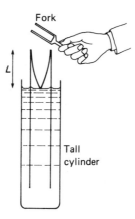

Figure 4.11

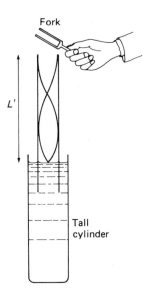

Figure 4.12

The tube is clamped to a stand and so its position can be adjusted. A tuning fork of *known frequency f* is sounded over the tube when it is almost totally immersed in water and as the tube is raised slowly, the intensity of the sound produced reaches a maximum—when **stationary wave resonance** is occurring.

The **fundamental** of the air column equals the frequency of the fork, and if L is the height of the air column above the water surface and c is the end correction for the tube, then the wavelength of the note produced is given by

$$\lambda/4 = L + c$$

If the fork is sounded again and the tube is raised even higher, a second, weaker resonance is found, and the frequency of vibration of the air column is still equal to the tuning fork frequency. Now the wavelength of the note emitted is given by

$$3\lambda/4 = L' + c$$

where L' is the new height of the air column.

The two equations can be used to eliminate c, giving

$$L' - L = \lambda/2$$

and from the wave equation

$$\lambda = v/f$$

then

$$\boxed{v = 2(L' - L)f}$$

An alternative method just involves making the first measurement as above, but doing it for a number of tuning forks with known frequencies f. Thus, once again

$$\lambda/4 = L + c$$

and so
$$\lambda = 4(L+c)$$

and by applying the wave equation,
$$v = 4f(L+c)$$

Therefore

$$L = \frac{v}{4f} - c$$

This is of the form $y = mx + c$ and if a graph of L versus $1/f$ is plotted, a straight line should result. The gradient of this line gives $v/4$ and the y intercept will give the end correction.

Example 4.1

A plane progressive sound wave strikes a reflector with zero angle of incidence. If it has a wavelength of 1.8 m, what are the amplitude and pressure changes at (i) 45 cm, (ii) 90 cm and (iii) 15 cm from the point of reflection?

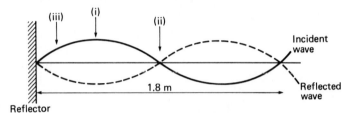

Figure 4.13

Figure 4.13 shows the displacements for both incident and reflected waves.

(i)   At 45 cm from the reflector (quarter of a wavelength), an *antinode* will occur. The molecules of the medium there will be further apart than usual and so the pressure will be a *minimum*.

(ii)   90 cm (half a wavelength) from the reflector, a *node* will arise and the molecules will be closest together. Thus, the pressure there will be a *maximum*.

(iii)   At 15 cm from the reflector the displacement of each wave is

$$y = \sin(15 \times 360°/180) = \sin 30° = 0.5$$

Thus, the resultant amplitude here will be *half the maximum* amplitude and the pressure is similarly *half the maximum* value.

Example 4.2

Ȧ   Ḃ   Ċ   Ḋ   Ė   Ḟ   Ġ   (A, B, C, D, E, F and G are seismographs)

X̣   Y̧   (X and Y are epicentres)

Circular shock waves of equal frequency, amplitude and phase arise from two epicentres X and Y. (i) Determine the wavelengths of the shock waves if maximum disturbances are recorded by the seismographs B and D but a minimum disturbance is recorded by C. (ii) Explain what would happen if the epicentres are brought closer together.

(i) For a minimum disturbance to occur at C, the waves from X and Y must arrive 180° out of phase, i.e. the difference in the distances travelled by the waves will be half a wavelength.

$$CY - CX = \lambda/2$$

For constructive interference to occur at B, the waves from X and Y must arrive in phase at B, i.e. the difference in distances travelled by the waves is one whole wavelength, assuming that B is the first maximum disturbance away from D.

$$BY - BX = \underline{\lambda}$$

(ii) If the epicentres were closer together, then the separation of the maximum and minimum disturbances, e.g. between B and C, would be larger.

Example 4.3

A glass tube is placed vertically so that the bottom part of it stands in water. It has an end correction of 0.01 m and the air temperature when it is made to produce its fundamental is 20 °C. Find the length of tube above the water surface if it is to produce a note which has a frequency of 520 Hz. (The speed of sound at 0 °C is 330 m s^{-1}.)

First, the speed of sound at 20 °C must be found. The speed of sound is directly proportional to the square root of the absolute temperature of the air. If the constant of proportionality is k, then

$$v_1 = k\sqrt{T_1} \quad \text{and} \quad v_2 = k\sqrt{T_2}$$

Therefore

$$\frac{v_1}{\sqrt{T_1}} = \frac{v_2}{\sqrt{T_2}} \quad \text{and} \quad v_2 = v_1\sqrt{\frac{T_2}{T_1}}$$

Thus, the speed of sound at 20 °C is

$$v_2 = 330\sqrt{\frac{293}{273}} = 342 \, \text{m s}^{-1}$$

The wavelength of the sound produced at 20 °C is given by

$$\lambda = v_2/f = 342/520 = 0.66 \, \text{m}$$

For the fundamental of a closed tube, $\lambda/4 = L + c$, where L and c are the length and end correction for the tube respectively. Thus,

$$L = \lambda/4 - c = \frac{0.66}{4} - 0.01 = \underline{0.16 \, \text{m}}$$

Example 4.4

The tube used in example 4.3 is once again set up so that 0.16 m of it is above the surface of the water. Determine the frequency of beats produced if a small speaker emitting a note of frequency 520 Hz is placed close to the tube when it is made to sound its fundamental at a temperature for the air of 30 °C.

As the temperature of the air has changed, then the new speed of sound must be found:

$$v_3 = v_1\sqrt{\frac{T_3}{T_1}} = 330\sqrt{\frac{303}{273}} = 348 \, \text{m s}^{-1}$$

Thus, the frequency of vibration for the tube is given by

$$f = v_3/\lambda = 348/0.66 = 527 \text{ Hz}$$

Therefore the frequency of beats

$$f_b = 527 - 520 = \underline{7 \text{ Hz}}$$

Example 4.5

Find the fundamental frequency produced by a string fixed at both ends if its length is 1.5 m and it is under a tension of 100 N. (The mass of the string is 0.3 mg.)

The fundamental is given by

$$f = \frac{1}{\lambda}\sqrt{\frac{T}{m/L}} = \frac{1}{2L}\sqrt{\frac{T}{m/L}}$$

Therefore

$$f = \frac{1}{3.0}\sqrt{\frac{100}{0.3 \times 10^{-6}/1.5}} = \underline{7.45 \text{ kHz}}$$

Doppler effect

This is the *apparent change* in the frequency of a wave motion when there is relative motion between source and observer.

Source moving

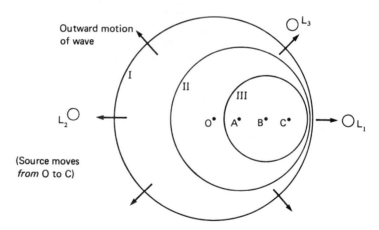

Figure 4.14

Figure 4.14 shows a source moving from O to C with a speed u emitting waves of speed v and frequency f. This is a 'snapshot' of the waves giving their positions when the source has reached C. Waves I, II and III were emitted at O, A and B respectively.

To the observer L_1 the waves are crowded together as the source advances *towards* him.

L_2 sees the waves spread out as the source goes *away* from him. The *apparent* wavelength in either case is given by the following:

$$\text{apparent wavelength} = \frac{\text{distance from source of waves}}{\text{number of waves in that distance}}$$

Thus, the respective apparent wavelengths for L_1 and L_2 are

$$\lambda_1 = \frac{v-u}{f} \quad \text{and} \quad \lambda_2 = \frac{v+u}{f}$$

Since the speed of the waves is unchanged, the *apparent frequency* as noted by each observer is then

$$f_1 = \frac{v}{(v-u)/f} \quad \text{and} \quad f_2 = \frac{v}{(v+u)/f}$$

As the source moves past the observer L_3 the change in frequency is $f_1 - f_2$ and there is a *change in pitch*.

Observer moving

If the source is stationary, the wavelength is constant. When the observer moves *towards* the source, the waves come to him faster than usual, with speed $v + u_L$. When he moves *away*, they appear to be slower than usual, with speed $v - u_L$. (u_L is the observer speed.)

Thus, the apparent frequencies for observer approach and recession, f_a and f_r respectively, are

$$f_a = (v + u_L)f/v \quad \text{and} \quad f_r = (v - u_L)f/v$$

Example 4.6

A source emits a continuous note of frequency 150 Hz. Determine the apparent frequency experienced by a listener if: (i) the listener does not move but the source approaches at 7 m s^{-1}; (ii) the source does not move but the listener approaches at 11 m s^{-1}; (iii) both source and listener approach each other at 7 m s^{-1} and 11 m s^{-1} respectively. (Assume speed of sound is 320 m s^{-1}.)

(i) The apparent frequency of the note is due to the apparent change in wavelength as the source approaches the listener.

$$\lambda_a = \frac{v - u_s}{f}$$

Thus,

$$f_a = v/\lambda_a$$

and so

$$f_a = \frac{vf}{v - u_s} = \frac{320 \times 150}{320 - 7} = \underline{153 \text{ Hz}}$$

(ii) The apparent frequency of the note is due to the apparent wave velocity as the listener approaches the source.

$$v_a = v + u_L$$

Thus,

$$f_a = v_a/\lambda$$

and so

$$f_a = \frac{v + u_L}{v/f} = \frac{(v + u_L)f}{v} = \frac{(320 + 11)150}{320} = \underline{155 \text{ Hz}}$$

(iii)   Since both source and listener are moving, the apparent frequency of the note is due to a combination of an apparent change in both wavelength and wave velocity. Thus,

$$f_a = \frac{v_a}{\lambda_a} = \frac{(v + u_L)f}{v - u_s} = \frac{(320 + 11)150}{320 - 7} = \underline{159 \text{ Hz}}$$

Example 4.7

A jet fighter flies in a straight line over an airfield at a speed of 264 m s^{-1} level with the ground, at a height of 600 m. If it emits a continuous note of frequency 1 kHz, what is the change in the apparent frequency as experienced by one of the ground crew over a 1.5 s interval as the jet flies overhead? (Assume speed of sound is 340 m s^{-1}.)

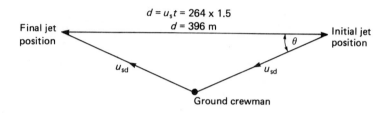

Figure 4.15

See Fig. 4.15. The component of velocity of the jet in the direction of the member of the ground crew is

$$u_{sd} = u_s \cos \theta = 264 \times \cos 71.7° = 83 \text{ m s}^{-1}$$

Thus, frequency range is

$$\frac{vf}{v - u_{sd}} - \frac{vf}{v + u_{sd}} = \underline{519 \text{ Hz}}$$

5
Light

Introduction

This chapter is yet another extension to the material covered in the chapter on waves. It begins with a comparison of the two theories concerning the nature of light, i.e. the corpuscular and wave theories. The idea of reflection is then used to introduce the two types of image which can be formed by an optical system.

The following section deals with the theory behind the prism and the deviation of light rays through it. This leads to the derivation of the equation for minimum deviation in a prism, and a description of dispersion. It is through dispersion that a prism is able to form a spectrum.

In the next part of the chapter the action of thin lenses is investigated and this is supported by ray diagram treatment of the formation of images using objects at different distances from the lens. Reference is then made to the use of a sign convention in the calculation of image position by application of the lens formula.

The definition of the linear magnification of a lens and experimental determination of the focal lengths of both converging and diverging lenses conclude the chapter.

Revision targets

The student should be able to:

(1) Discuss the nature and basic properties of light waves.
(2) Define the terms real image and virtual image.
(3) Define and give the equation for the total deviation of light passing through a prism.
(4) Explain when minimum deviation occurs in a prism and use it to calculate the refractive index for a prism material.
(5) Describe how a prism is able to split up white light into the constituent colours of the visible spectrum.
(6) Define the term principal focus and explain what is meant by the focal length of a lens.
(7) Represent the images formed by a lens graphically and apply the lens formula to deduce their nature.
(8) Define the term magnification for a lens.
(9) Describe an accurate method for the determination of the focal length of a converging lens.
(10) Explain how a converging lens can be used to determine the focal length of a diverging lens.
(11) Explain what is meant by a virtual object.

Corpuscular theory

Corpuscular theory, supported by **Newton**, considered light to consist of tiny particles travelling in straight lines.

Wave theory

Wave theory, proposed by **Huygens**, considered light to be of progressive or travelling wave form.

Properties of light

Property	Predicted by:		Significant property of other electromagnetic wave
	Corpuscular theory	Wave theory	
Rectilinear propagation	Yes	Yes	All
Reflection	Yes	Yes	Radio
Refraction	Yes	Yes	Infrared
Light travels faster in air than water	No	Yes	—
Photoelectric effect	Yes	No	Ultraviolet
*Compton effect	Yes	No	X-rays
Interference	No	Yes	Radio
Diffraction	No	Yes	X-rays
Energy transfer	Yes	Yes	All

*Compton effect: Effect in particle physics where the wavelength of radiation changes as a result of an interaction with a free electron.

Maxwell showed that light was made up of oscillating electric and magnetic fields and hence did not need a medium for transmission as required by the wave theory.

Duality of light

At present it is thought that light has a *dual nature*. The phenomena of light propagation and the properties of long wavelength electromagnetic radiation (radio, infrared, etc.) are easily explained by the wave theory. The interaction of light with matter and the properties of short wavelength electromagnetic radiation (X-rays, ultraviolet, etc.) are better explained by the corpuscular theory.

Real image

A **real** image is an image formed by a mirror or lens at a point through which rays of light *actually* pass. The image can be obtained on a screen.

Virtual image

A **virtual** image is seen at a point from which rays of light appear to come. It cannot be put on a screen.

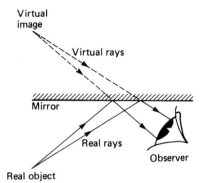

Figure 5.1

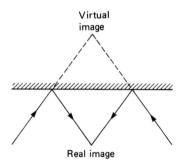

Figure 5.2

A plane mirror normally gives a *virtual image*, but can be made to form a real image by using a convergent beam of light as shown in Figs 5.1 and 5.2.

Refraction through prisms

Angle A in Fig. 5.3 is called the **refracting angle**, or simply **the angle**, of the prism, and QP is the **refracting edge**.

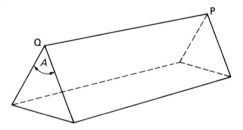

Figure 5.3

In Fig. 5.4:

for refraction at EF, the angle of deviation $= i_1 - r_1$

for refraction at EG, the angle of deviation $= i_2 - r_2$

Therefore the **total deviation** D is given by:

$$D = (i_1 - r_1) + (i_2 - r_2)$$

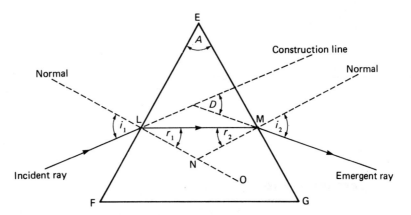

Figure 5.4

From the geometry of a quadrilateral, it can be shown that

$$\angle MNO = A$$

but, since $\angle MNO = r_1 + r_2$ from exterior angle theory, then

$$\boxed{A = r_1 + r_2}$$

The variation of the angle of incidence i_1 with deviation D (Fig. 5.5) shows a minimum value for deviation. At this point light passes *symmetrically* through the prism.

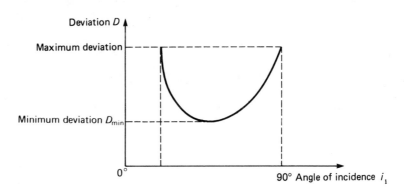

Figure 5.5

Minimum deviation

The minimum deviation D_{min} in a prism occurs when $i_1 = i_2$ and $r_1 = r_2$. Let i and r equal these angles respectively. Then

$$D_{min} = 2(i - r) \quad \text{and} \quad A = 2r$$

Thus,

$$D_{min} = 2i - A$$

Rearranging,

$$i = \frac{A + D_{min}}{2}$$

Finally, if n is the refractive index for the material of the prism, then from *Snell's law of refraction*:

$$n = \frac{\sin i}{\sin r}$$

Thus,

$$n = \frac{\sin\left(\dfrac{A + D_{\min}}{2}\right)}{\sin(A/2)}$$

Dispersion due to a prism

When a prism splits up white light into colours of the spectrum, violet light is deviated *most* while red is deviated *least*. The *refraction* and therefore the *refractive index* for violet light is greater than it is for red light.

A pure spectrum of *non-overlapping* colours is obtained by passing a parallel beam of white light through a prism (Fig. 5.6). A plano-convex lens is used to produce the pure spectrum from the emergent beam.

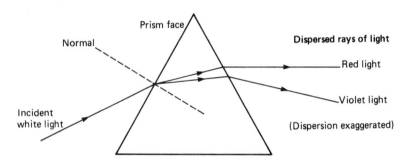

Figure 5.6

Thin lenses

There are two main types of thin lens: **convex** lenses which *converge* paraxial rays of light, and **concave** lenses which *diverge* paraxial rays. A **paraxial ray** is one which is close to and parallel with the principal axis. This is the line joining the centres of curvature of the lens surfaces.

A **principal focus** is a point on the principal axis to which paraxial rays converge (converging lens), or from which they appear to diverge (diverging lens), after passing through the lens (Figs 5.7 and 5.8). Lenses

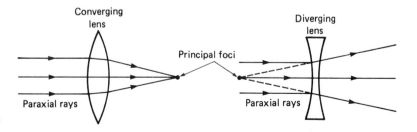

Figure 5.7   Figure 5.8

have two principal foci because of their two surfaces and the distance of each from the centre of the lens is called the **focal length**.

Example 5.1

A prism is made of plastic which has refractive index of 1.39. It is placed into a non-reactive liquid of refractive index 1.29 and light is then shone at it. Determine the minimum deviation D^* for light being refracted through the prism if the refracting angle of the prism is $62°$.

Using the subscripts l and p for the liquid and prism respectively, then:

$$_l n_p = \frac{\sin\left(\dfrac{A+D^*}{2}\right)}{\sin(A/2)}$$

But

$$_l n_p = \frac{_{air} n_p}{_{air} n_l}$$

and so

$$\sin\left(\frac{A+D^*}{2}\right) = \frac{_{air} n_p}{_{air} n_l} \times \sin\left(\frac{A}{2}\right)$$

Therefore

$$\sin\left(\frac{A+D^*}{2}\right) = \frac{1.39}{1.29} \times \sin\left(\frac{62°}{2}\right) = 0.555$$

Thus,

$$\frac{A+D^*}{2} = 33.7°$$

So

$$D^* = (33.7 \times 2) - 62$$

Therefore

$$D^* = \underline{5.4°}$$

Example 5.2

The prism in example 5.1 is now removed from the liquid and placed in the open air. Find (i) the angle of incidence which will produce a maximum deviation, and (ii) the greatest prism angle which would permit refraction if the prism were to be made of the same material.

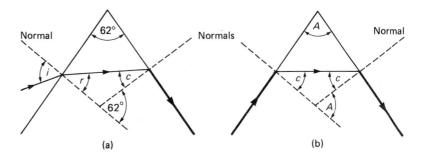

Figure 5.9

(a)   (b)

(i)   Maximum deviation will occur when the ray of light emerges from the second prism face along the surface (see Fig. 5.9(a)).

The angle c is thus the critical angle. Now,

$$\sin c = \frac{1}{{}_{air}n_p} = \frac{1}{1.39} = 0.719$$

Thus,

$$c = 46°$$

But

$$r = 180° - c - (180° - 62°) = -c + 62° = 16°$$

Now,

$$\sin i = {}_{air}n_p \sin r = 1.39 \times \sin(16°) = 0.383$$

and $\quad i = \underline{22.5°}$

(ii) Since the refracting angle of the prism $A = r_1 + r_2$, then the greatest value of A will occur when light is incident and emerges along the prism faces, i.e. both the refracting angles are equal to the critical angle of the prism material. Therefore

$$A = c + c = 2c = 2 \times 46° = \underline{92°}$$

(See Fig. 5.9(b)).

Ray diagrams

Two of the following rays drawn from the top of the object form the image on a **ray diagram**, which is a graphical representation of what occurs to light when it passes through a lens.

(1) A *paraxial ray* which after refraction passes through or appears to diverge from the principal focus.

(2) A ray through the *optical centre* of the lens which is unaltered.

(3) A ray through the *principal focus* refracted parallel to the principal axis.

Real is positive convention

It is possible to *calculate* the position of an image without using ray diagrams. To establish the *nature* of the image a sign convention is required. There are a number of sign conventions, and the one used here is called the **real is positive convention**. The symbols employed are as follows.

u = distance of the object from the centre of the lens

v = distance of the image from the centre of the lens

f = focal length of lens

The following rules need to be applied when using the sign convention.

(1) Distances of *real* objects and images are *positive*.

(2) Distances of *virtual* objects and images are *negative*.

(3) The focal length of a converging lens is *positive*, while the focal length of a diverging lens is always *negative*.

The calculations are based on the **lens formula**:

$$\boxed{\frac{1}{f} = \frac{1}{u} + \frac{1}{v}}$$

Ray diagrams for lenses

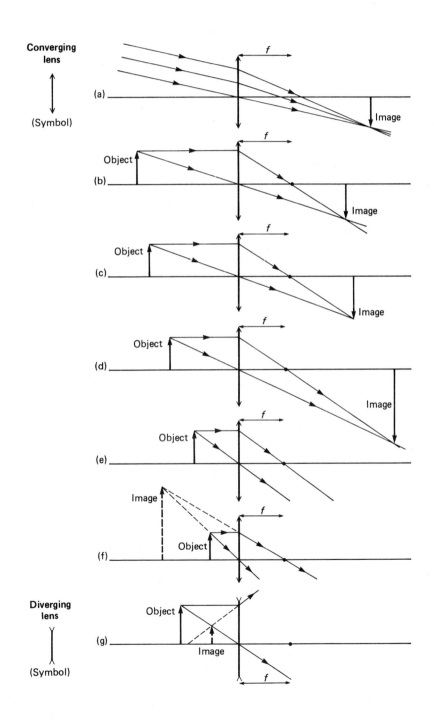

Figure 5.10

The following table is the key to the ray diagrams shown in Fig. 5.10.

Diagram	Object position	Image position	Nature of image	Application
a	Infinity	Principal focus	Real, inverted, smaller	Telescope objective
b	Between ∞ and 2f	Between f and 2f	Real, inverted, smaller	Eye lens
c	At 2f	At 2f	Real, inverted, same size as object	—
d	Between 2f and f	Between 2f and ∞	Real, inverted, enlarged	Cine projector
e	At f	At ∞	Virtual, erect, enlarged	Spotlight lens
f	Between f and lens	Behind object	Virtual, erect, enlarged	Magnifier
g	Any	In front of object	Virtual, erect, smaller	Telescope eyepiece

Magnification due to a lens

All paraxial rays from a point object after refraction form a point image. However, the magnification for an image of an object which is not a point is

$$m = \frac{\text{height of image}}{\text{height of object}} = \frac{v}{u}$$

Other lens formulae

(1) For two lenses in contact of focal lengths f_1 and f_2, the focal length of the combination f is given by

$$\frac{1}{f} = \frac{1}{f_1} + \frac{1}{f_2}$$

where the sign convention is applied.

(2) The power of a lens is the reciprocal of the focal length in metres.

$$F = \frac{1}{f}$$

(The unit of optical power is the dioptre.)

Converging lenses have positive F, while diverging lenses have negative F.

Measurement of focal length

Converging lenses

(a) **Plane mirror method** Here a converging (convex) lens is placed upon a plane mirror (Fig. 5.11). A pin held parallel to the plane mirror,

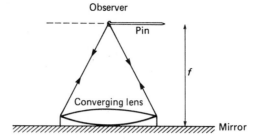

Figure 5.11

above the lens, is moved up and down until it coincides with its inverted image. The distance to the lens is then the focal length of the lens.

(b) **Lens formula method** This is a more accurate method than (a). Values of object and image distance are measured using either the pin no-parallax method or an illuminated object and screen. Values of $1/u$ and $1/v$ are then plotted on a graph. From the lens formula the *intercept* on each axis should be $1/f$.

(c) **The displacement method** This depends on an important concept. It is called the *reversibility of light*, and means that if the direction of light is reversed it will take the same path it came along (Fig. 5.12).

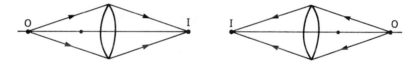

Figure 5.12

O and I are called **conjugate points**, because they are interchangeable.

The image I of an illuminated object O is obtained on a screen. With the object and the screen in the same position the lens is moved until another sharp image is formed. A formula then gives f.

Real image formation in a converging lens

A converging lens *cannot* form a real image if (i) the object is less than the focal length from the lens and (ii) the distance between the conjugate points is less than four times the focal length. (The latter can be shown by a mathematical treatment.)

Focal length of the diverging lens

A practical problem with the diverging lens is that it forms only **virtual** images which cannot be produced onto a screen. To overcome this a **converging** lens is employed.

The converging lens forms a real image I_2 of an object O, when the diverging lens is not present.

Using a screen this image is located and its position noted. The diverging lens is now placed between the converging lens and the real image (Fig. 5.13).

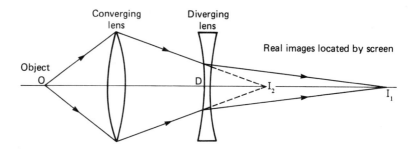

Figure 5.13

This real image I_2 behaves as a **virtual object** for the diverging lens and produces a **real** image I_1. The new image is located and its position once again noted.

For the diverging lens $u = -I_2D$ and $v = +I_1D$. Thus, applying the lens formula, the focal length of the diverging lens f is given by

$$\boxed{\frac{1}{f} = \frac{1}{-I_2D} + \frac{1}{I_1D}}$$

Example 5.3

An object is placed 0.2 m away from a diverging lens which has focal length 0.15 m. By application of the lens formula, determine (i) the position of the image and (ii) the nature and magnification of the image.

(i)   Applying the lens formula,

$$\frac{1}{f} = \frac{1}{u} + \frac{1}{v}$$

Then

$$\frac{1}{v} = \frac{1}{f} - \frac{1}{u} = \frac{1}{-0.15} - \frac{1}{0.2} = -11.67 \text{ m}^{-1}$$

Thus,

$$v = -0.086 \text{ m}$$

(ii)   From the real is positive sign convention, since the image distance is negative, then the image is *virtual*.

The linear magnification of the image is $m = v/u$. Therefore

$$m = 0.086/0.2 = \underline{0.43}$$

Hence the image is diminished, i.e. smaller than the object. Note also that when using the equation for linear magnification the sign of the image and object distances is ignored.

Example 5.4

A converging lens of focal length 20 cm is in contact with a diverging lens of focal length 25 cm. Find the position of the image of an object placed 15 cm away from this combination.

The resultant focal length of the combination is given by

$$\frac{1}{f} = \frac{1}{f_1} + \frac{1}{f_2} = \frac{1}{-0.25} + \frac{1}{0.2} = 1.0 \text{ m}^{-1}$$

Thus, $f = +1$ m (i.e. the combination acts like a converging lens).

Now, applying the lens formula,

$$\frac{1}{f} = \frac{1}{u} + \frac{1}{v}$$

then

$$\frac{1}{v} = \frac{1}{f} - \frac{1}{u} = \frac{1}{1} - \frac{1}{0.15} = -5.67 \, \text{m}^{-1}$$

Therefore

$$v = -0.18 \, \text{m}$$

(The image formed is *virtual*.)

Example 5.5

An object is placed 0.8 m in front of a converging lens of focal length 0.5 m. Another converging lens of focal length 0.2 m lies a distance of 1.0 m behind the first lens. Determine the position of the image and its nature.

For the first converging lens,

$$\frac{1}{f} = \frac{1}{v} + \frac{1}{u}$$

Therefore

$$\frac{1}{v} = \frac{1}{f} - \frac{1}{u} = \frac{1}{0.5} - \frac{1}{0.8} = 0.75 \, \text{m}^{-1}$$

Thus,

$$v = 1.33 \, \text{m}$$

Hence for the second converging lens this image formed by the first one acts like a virtual object.

So for the second lens,

$$u = -(1.33 - 1.0) = -0.33 \, \text{m}$$

and so

$$\frac{1}{v} = \frac{1}{f} - \frac{1}{u} = \frac{1}{0.2} - \frac{1}{-0.33} = 8.03 \, \text{m}^{-1}$$

Therefore

$$v = 0.13 \, \text{m}$$

Thus, the image formed by the combination is real and lies 0.13 m behind the second lens, or 1.13 m behind the first lens.

Example 5.6

A converging lens of focal length 45 cm is placed in contact with another lens X. If this combination of lenses has a power of +5 dioptre, determine the focal length and power of lens X.

For the combination of lenses,

$$\frac{1}{f} = \frac{1}{f_1} + \frac{1}{f_2}$$

Also, the power of the combination is given by $F = 1/f$. Therefore

$$F = \frac{1}{f_1} + \frac{1}{f_2}$$

and so

$$\frac{1}{f_2} = F - \frac{1}{f_1}$$

Thus,

$$\frac{1}{f_2} = 5 - \frac{1}{0.45} = 2.78 \, \mathrm{m}^{-1}$$

Therefore

$$f_2 = \underline{0.36 \, \mathrm{m}}$$

The power of this lens is given by

$$1/f_2 = \underline{2.78 \, \mathrm{dioptre}}$$

6
Heat

Introduction

This chapter deals with a most basic form of energy—heat—and begins by looking at the determination of heat gained by a body and the effect of heat loss. The ideal gas is then introduced and the laws governing its behaviour are investigated. With the aid of the kinetic theory, the equation for the pressure of such a gas is derived. This is a relationship between a macroscopic property—pressure—and a microscopic one— rms speed of the gas molecules. This is extended by looking at the connection between average molecular kinetic energy and the gas temperature.

Vapours and real gases are dealt with next, and the idea of latent heat is put in terms of the kinetic theory. A consideration of temperature, temperature scales, and types of thermometer leads to the final topic on the conduction of heat through materials.

Revision targets

The student should be able to:

(1) Define and apply specific heat capacity, to numerical problems.
(2) Describe the determination of specific heat capacity and discuss how the continuous flow method deals with heat loss.
(3) Explain what is meant by the kinetic theory and an ideal gas.
(4) State and apply the gas laws to numerical problems.
(5) Use the kinetic theory to obtain an equation for pressure in terms of the rms speed of the gas molecules.
(6) Distinguish between an average and an rms value.
(7) Appreciate the relation between average molecular kinetic energy and gas temperature, and emphasise its significance with a consideration of Avogadro's hypothesis and Dalton's law.
(8) Explain what is meant by a vapour and distinguish between boiling and evaporation.
(9) Use Andrews' experiment to outline real gas properties.
(10) Define critical temperature.
(11) Define the principal specific heat capacities of a gas.
(12) Use the idea of external work to explain why a gas can have an infinite number of specific heat capacities.
(13) Apply the kinetic theory to a discussion of the states of matter.
(14) Define the specific latent heats of fusion and vaporisation respectively.
(15) Discuss the first law of thermodynamics in terms of internal energy and use the law to explain the effects of various processes on a gas.

(16)   Explain the term temperature and distinguish it from heat.
(17)   Discuss the setting up of a temperature scale and give examples of such a scale.
(18)   Apply the relationship between temperature in °C and a thermometric property to the solution of numerical problems.
(19)   Describe the characteristics of various thermometers including liquid-in-glass, constant volume gas and resistance thermometers.
(20)   Explain the working of a thermocouple and the term neutral temperature.
(21)   Compare heat conduction in metals and non-metals.
(22)   Define coefficient of thermal conductivity and use it in the solution of problems involving conduction through composite slabs.
(23)   Describe the determination of the thermal conductivity for both good and bad conductors.

What is heat?

Heat is energy which is transferred by **convection**, **radiation** or **conduction** from one body to another at a lower temperature.

Different materials of the same mass are heated. For the same temperature rise they need different amounts of heat. This property is measured in terms of **heat capacity**.

Heat capacity (symbol *C*)

The heat energy needed to raise the temperature of a material by one *kelvin* (units: $J\,K^{-1}$).

Specific heat capacity (symbol *c*)

The heat energy needed to raise the temperature of 1 kg of a material by one kelvin (units: $J\,kg^{-1}\,K^{-1}$).

Examples

Material	$c\ (J\,kg^{-1}\,K^{-1})$
Water	4185
Methylated spirits	2500
Aluminium	920
Copper	385

Heat equation

heat lost = heat gained

Since specific heat capacity is defined in terms of temperature change and mass, heat energy can be obtained from the following:

$$\text{heat energy} = \frac{\text{mass of}}{\text{substance}} \times \frac{\text{specific heat capacity}}{\text{of substance}} \times \frac{\text{temperature}}{\text{change}}$$

In symbols this is often written as

$$Q = mc\Delta\theta$$

Example 6.1

How much heat is given out when an iron poker of mass 0.5 kg cools from 70 °C to 10 °C? (Specific heat capacity of iron is 460 J kg^{-1} K^{-1}.)

temperature change $\Delta\theta = 60$ K

Therefore applying

$$Q = mc\Delta\theta$$
$$Q = 0.5 \times 460 \times 60 = \underline{13\,800\,J}$$

Measuring the specific heat capacity of a metal

A metal block of known mass m is lagged with expanded polystyrene, cotton wool or felt to reduce heat losses to the air. A thermometer and immersion heater are then put into holes in the block using oil or mercury to ensure good thermal contact. See Fig. 6.1.

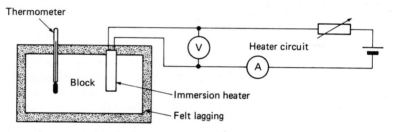

Figure 6.1

The initial temperature of the block is noted and it is heated for a time t with a current I through the immersion heater and a potential difference V across it. The final temperature is also recorded.

From the heat equation,

$$Q = mc\,\Delta\theta = VIt$$

Therefore

$$c = \frac{VIt}{m\,\Delta\theta}$$

Specific heat capacity c_L of a liquid by electrical heating

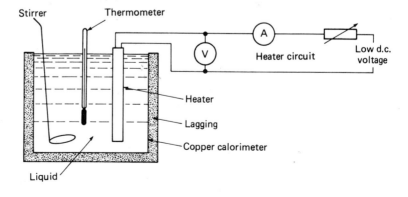

Figure 6.2

The mass of a copper calorimeter when empty m_c and when full of liquid $m_c + m_L$ is found.

The liquid is heated using the circuit in Fig. 6.2 and constantly stirred. The current I and potential difference V and time t for heating are all recorded, along with the initial and final liquid temperatures T_i and T_f respectively.

$$\text{temperature rise} = T_f - T_i = \Delta\theta$$

From the heat equation,

$$VIt = m_L c_L \Delta\theta$$

from which c_L can be obtained.

However, some of the heat has gone into warming up the calorimeter. Call this heat H. Then

$$\boxed{VIt = m_L c_L \Delta\theta + H}$$

where $H = m_c c_c \Delta\theta$.

Method of mixtures

This can be used to find the specific heat capacity of a solid c_s. It is a good method for *poor conductors* of heat unlike aluminium or copper.

A sample of the substance is heated for some time in a beaker of boiling water (373 K). It is then transferred quickly to a lagged copper calorimeter of mass m_c containing water of mass m_w at a known initial temperature T_i. (The sample has a thread attached for easy transferrence to the calorimeter.)

After *continuous stirring* the final steady temperature T_f is noted.

$$\text{heat given out by sample} = m_s c_s (373 - T_f)$$

$$\text{heat gained by water in the calorimeter} = m_w c_w (T_f - T_i)$$

From the heat equation,

$$m_s c_s (373 - T_f) = m_w c_w (T_f - T_i)$$

Therefore

$$c_s = \frac{m_w c_w (T_f - T_i)}{m_s (373 - T_f)}$$

Heat loss $H = m_c c_c (T_f - T_i)$ will occur to the calorimeter and this can be compensated for by inclusion of H in the above equations.

Importance of specific heat capacity

(1)   To ensure good thermal contact between surfaces, a liquid such as oil or mercury is used. These are *good conductors* of heat and have fairly low specific heat capacities.

(2)   Water has a *high* specific heat capacity and therefore can absorb a lot of heat before its temperature goes up by one kelvin. Hence its action as a coolant.

(3)   Water will also give out a lot of heat when its temperature goes down by one kelvin and therefore it is used in radiators, etc.

Notes on calorimeter experiments concerning heat loss

(1) High polishing of calorimeter required to stop radiation (very small).
(2) Outer container required to stop convection and conduction.
(3) Insulation required between calorimeter and outer container to stop conduction. (As required by domestic hot water storage tanks, for example.)
(4) Continuous stirring of liquid required during rise in temperature.
(5) Final liquid temperatures need to be *corrected for cooling*.

Cooling correction

The procedure for this is to measure the temperature rise every 30 s and continue with this until the liquid has cooled from its final temperature by at least one kelvin. A graph of temperature versus time is then plotted, as shown in Fig. 6.3. (T_0 is the initial temperature from which timing was begun, and A_1 and A_2 are areas.)

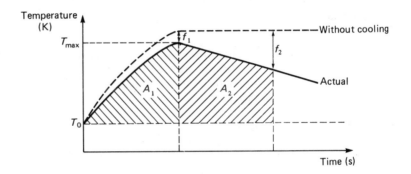

Figure 6.3

From *Newton's cooling law*, the rate of loss of heat is directly proportional to the temperature difference between the body and the surroundings. This can be shown in the following expression:

$$f_1 = f_2 \frac{A_1}{A_2}$$

Then the corrected maximum temperature is $f_1 + T_{max}$ where f_2 has been taken as one kelvin.

This cooling correction is only applicable when there is heat loss by conduction or by convection when the cooling body is in a draught. It is also true for convection when the body is in 'still' air and for radiation, so long as the temperature differences involved are small.

Callendar and Barnes method

This is a continuous flow method for finding the specific heat capacity of water c_w.

Precautions

Before and during setting up the apparatus, water is allowed to pass through the tap supplying the constant head at a fast rate for about ten minutes. This is done in order to obtain a constant temperature in the water, since temperature differences exist between water in pipes and tank, etc. As a lot of wiring is involved all connections are checked.

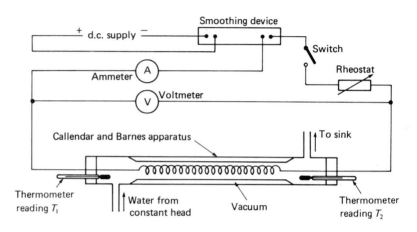

Figure 6.4

The apparatus is set up as shown in Fig. 6.4. The variable resistor and d.c. supply are set to give a fairly high current, and the rate of flow of water through the Callendar and Barnes apparatus adjusted to a fairly fast flow.

Using a beaker and balance the mass of water m_1 collected over a time t is found.

When T_1 and T_2 are stable (i.e. *steady state*) their values along with those of current I_1 and potential difference V_1 are noted. The experiment is now repeated using a value of current about half of the original and the rate of flow re-adjusted to maintain the temperature T_2. The temperature of the incoming water should remain the same.

By application of the heat equation,

heat supplied = heat absorbed by water + heat lost

So $V_1 I_1 t = m_1 c_w (T_2 - T_1)$ $+ H$
and $V_2 I_2 t = m_2 c_w (T_2 - T_1)$ $+ H$

By subtraction, the heat loss H can be *eliminated*. So

$$V_1 I_1 t - V_2 I_2 t = (m_1 - m_2) c_w (T_2 - T_1)$$

Therefore

$$c_w = \frac{(V_1 I_1 - V_2 I_2)t}{(m_1 - m_2)(T_2 - T_1)}$$

For small temperature difference $T_2 - T_1$, the specific heat capacity of water at $(T_2 - T_1)/2$ is obtained.

Example 6.2

An electrical heater using a current of 2.7 A is put into a calorimeter of mass 0.15 kg containing 160 g of a liquid Y. Find the potential difference across the heater if the temperature of Y rises by 100 °C in 950 s. The specific heat capacities of Y and the calorimeter material are $3 \times 10^3 \, \text{J kg}^{-1} \text{K}^{-1}$ and $2 \times 10^3 \, \text{J kg}^{-1} \text{K}^{-1}$ respectively.

heat supplied by heater = heat gained by Y and calorimeter

Thus, electrical energy

$$E = VIt = (m_y c_y + m_c c_c)(\theta_2 - \theta_1)$$

where V is the potential difference across the heater, I the current and t the time of heating. (Subscripts y and c have been used for the liquid and the calorimeter respectively.)

Therefore

$$V = (m_y c_y + m_c c_c)\frac{(\theta_2 - \theta_1)}{It}$$

and so

$$V = (480 + 300) \times \frac{100}{2.7 \times 950} = \underline{30 \, \text{V}}$$

Example 6.3

The following data were obtained during continuous flow calorimeter experiments on water and a liquid X.

Liquid	Flow (kg min^{-1})	Current (A)	p.d. (V)	Temperature rise (K)
Water	6×10^{-2}	4.5	5.2	5.0
X	18×10^{-2}	5.5	7.7	5.0

If the specific heat capacity of water is $4.18 \times 10^3 \, \text{J kg}^{-1} \text{K}^{-1}$ find (i) the rate at which heat is lost from the calorimeter and (ii) the specific heat capacity of the liquid X.

(i) For the water,

$$IVt = mc_w(\theta_2 - \theta_1) + H \qquad (H = \text{heat loss})$$

Therefore

$$\frac{H}{t} = IV - \frac{mc_w}{t}(\theta_2 - \theta_1)$$

Thus,

$$\frac{H}{t} = (4.5 \times 5.2) - \frac{(6 \times 10^{-2} \times 4180 \times 5.0)}{60}$$

Therefore

$$\frac{H}{t} = \underline{2.5 \, \text{W}}$$

(ii) For the water,

$$VIt = mc_w(\theta_2 - \theta_1) + H \qquad\qquad \text{equation (1)}$$

For X,

$$V^*I^*t = m^* c_X(\theta_2 - \theta_1) + H \qquad\qquad \text{equation (2)}$$

Subtracting equation (2) from equation (1) gives

$$(IV - I^*V^*)t = (mc_w - m^*c_X)(\theta_2 - \theta_1)$$

Thus,

$$mc_w - m^*c_X = \frac{(IV - I^*V^*)t}{(\theta_2 - \theta_1)}$$

and so

$$c_X = \frac{mc_w}{m^*} - \frac{(IV - I^*V^*)t}{(\theta_2 - \theta_1)m^*} = \left(\frac{6}{18} \times 4180\right) + 1263$$

Therefore

$$c_X = \underline{2.66 \times 10^3 \, \text{J kg}^{-1} \, \text{K}^{-1}}$$

Advantages of Callendar and Barnes method

As the temperatures are constant they can be measured very accurately with platinum resistance thermometers. Also the heat capacity of the apparatus is *not needed*. (No part of it is heated.) No cooling correction needs to be made as the repetition of the experiment for a different current value eliminates heat loss.

Ideal gases

When dealing with gases it is important to note that a *fixed mass* of gas is being dealt with. (This may not be so with vapours.)

An ideal (sometimes called **perfect**) gas is a *theoretical* gas which obeys the gas laws exactly. It has:

(1) perfectly elastic molecules,
(2) negligible intermolecular forces,
(3) negligible molecular volume.

The *gas laws* are hence true for real gases only at *low pressures* when the above conditions are approximately met.

Gas laws

Boyle's law

For a fixed mass of gas at constant temperature the product of pressure and volume is constant.

$$\boxed{PV = constant}$$

This has often been expressed as

$$P_1 V_1 = P_2 V_2$$

where P_1, P_2 and V_1, V_2 are different pressures and volumes respectively for the *same mass* of gas under *constant temperature*. In doing calculations on this law it is important to note that any units of pressure and volume can be used so long as they are the same on both sides of the above equation. (SI units of pressure and volume are Pa and m^3 respectively.)

The apparatus of Fig. 6.5 can be used to illustrate Boyle's law.

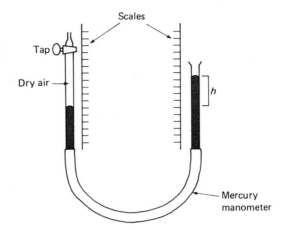

Figure 6.5

The pressure of the dry air is increased by increasing the height *h* and the volume noted from the scale. The mercury must be allowed to settle for the reading. *h* is also noted. The experiment is repeated and graphs drawn of *P versus V* and *P versus* $1/V$ (Fig. 6.6).

gas pressure = atmospheric pressure + *h*

Graphs for Boyle's law

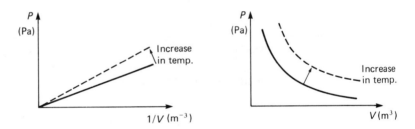

Figure 6.6

Charles' law

The volume of a fixed mass of gas increases per kelvin rise in temperature by a constant fraction of the volume at 273 K (0 °C), the pressure being constant throughout.

$$\frac{V}{T} = constant$$ (*T* is in *kelvin*)

The law can be demonstrated using the apparatus of Fig. 6.7.

Readings of column *height* and *temperature* are made making sure that at each reading the water is well *stirred*. The constant pressure applied to the air column is atmospheric pressure plus the pressure due to the mercury pellet. A plot of volume versus temperature in °C is shown in Fig. 6.8.

The lowest *theoretical* temperature is -273.15 °C. This is called *absolute zero* or the *zero of thermodynamic temperature*, 0 K (zero kelvin). A one kelvin change is equivalent to a one °C change.

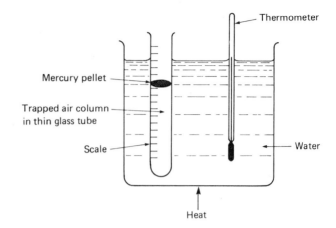

Figure 6.7

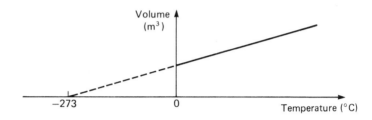

Figure 6.8

Pressure law

The pressure of a fixed mass of gas increases per kelvin rise in temperature by a constant fraction of the pressure at 273 K, the volume being constant throughout.

$$\frac{P}{T} = constant \qquad (T \text{ is in } kelvin)$$

Pressure law experiment

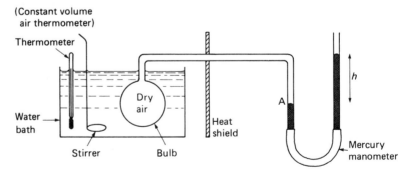

Figure 6.9

See Fig. 6.9. The dry air in the bulb is heated in the water bath. At intervals of temperature, the manometer is adjusted so that the mercury reaches level A. This keeps the volume of the gas *constant*. The height *h* is measured and the gas pressure equals H plus h, where H is the height of the mercury column due to atmospheric pressure. The temperature at each reading is also noted.

Kelvin temperature scale

Graphs of P versus T and V versus T for gases indicate a lowest or *absolute temperature* of $-273\,°C$ (see Fig. 6.10). All gases liquefy before this, but Kelvin used a *theoretical ideal gas* to define his scale of temperature. Accurate constant volume gas thermometers use hydrogen which is very near to being an ideal gas at low pressures.

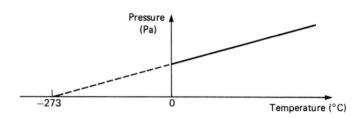

Figure 6.10

Gas equation

From the above laws the following expressions arise:

$$P \propto 1/V \quad \text{and} \quad P \propto T$$

For a *fixed mass* of gas these expressions can be combined to give:

$$\boxed{PV/T = constant}$$

This expression is called the **gas equation**.

Avogadro's hypothesis

Equal volumes of all gases under the same conditions of temperature and pressure contain in them equal numbers of molecules.

The mole

The mole is the amount of substance that contains as many elementary units as there are atoms in $0.012\,kg$ of carbon-12. (These units can be atoms, molecules, ions, electrons or particles.) The symbol for the mole is **mol**.

The number of *atoms* in $0.012\,kg$ of carbon-12 is called **Avogadro's constant**, N.

$$N = \underline{6.0 \times 10^{23}\,mol^{-1}}$$

For example:

one mole of electrons has a charge of $N \times$ charge on one electron

$$= 6 \times 10^{23} \times 1.6 \times 10^{-19} = \underline{96\,000\,C}$$

Ideal gas equation

For *one mole* of ideal gas, the constant in the gas equation is given the symbol R and is called the **universal gas constant**. It is 'universal' because it is the same for all gases. The equation

$$\boxed{PV/T = R}$$

is called the **ideal gas equation** or sometimes the equation of state for an ideal gas. As R is a constant it can be calculated by using just one set of conditions.

Standard temperature and pressure (stp)

Here the temperature is defined as $T = 273\,\mathrm{K}$ and the pressure as $P = 0.76\,\mathrm{m}$ of mercury. It is found from experiment that one mole of any gas at stp occupies a volume of

$$V = 0.0224\,\mathrm{m}^3$$

The units quoted above must be converted to SI. This is done by using force = mass × acceleration = density × volume × acceleration. So

$$\text{pressure} = \frac{\text{force}}{\text{area}}$$

$$= \frac{\text{density}}{\text{of Hg}} \times \frac{\text{height of}}{\text{Hg column}} \times \frac{\text{acceleration}}{\text{due to gravity}}$$

Thus, for *one mole* of gas,

$$R = \frac{PV}{T} = 13\,600 \times 0.76 \times 9.81 \times \frac{0.0224}{273}$$

Thus,

$$R = 8.3\,\mathrm{N\,m\,mol^{-1}\,K^{-1}} = \underline{8.3\,\mathrm{J\,mol^{-1}\,K^{-1}}}$$

For μ moles of gas,

$$\boxed{PV = \mu R T}$$

If the mass of the gas is m and the mass of one mole is M, then $\mu = m/M$. So

$$\boxed{PV = \frac{m}{M} RT}$$

M is called the **molar mass** and examples of it are given below.

Gas	M (kg)	M_r
Hydrogen	0.002	2
Nitrogen	0.028	28
Oxygen	0.032	32

Relative molecular mass (RMM)

This is defined by the following equation:

$$M_r = \frac{\text{mass of molecule}}{\text{mass of carbon-12 atom}} \times 12$$

Also, if there are n molecules of gas and since there are N molecules contained in one mole, then $\mu = n/N$ and so

$$PV = \frac{n}{N} RT$$

Kinetic theory

Particles of matter in all states of aggregation are in continual and random motion. These motions become more energetic with increase in temperature. This theory generally applies the laws of mechanics to gas molecules in an attempt to relate the microscopic properties of a gas to its macroscopic properties.

Macroscopic scale
Anything a magnifying glass is not required for observation of, e.g. a brick or planet or star, is said to be of macroscopic scale. Macroscopic properties include temperature and pressure.

Microscopic scale
Everything else is microscopic, e.g. molecules and atoms. Things on this sort of scale have properties such as speed or charge.

Pressure of an ideal gas

To apply the kinetic theory to an ideal gas the following assumptions as to the nature of the gas must be made.

(1)   The gas molecules behave as perfectly elastic spheres to which the laws of mechanics can be applied.
(2)   The molecules of the gas are in continual random motion.
(3)   There are so many molecules that averages can be used.
(4)   The interval of time during which an impact occurs is very small.
(5)   Intermolecular forces are negligible.
(6)   Molecular volumes are negligible.
(7)   Pressure of the gas is due to molecules bombarding the walls of any containing vessel.

Consider a cubic container of side L with n gas molecules each of mass m (Fig. 6.11).

Axes x, y and z are aligned as shown in Fig. 6.11. On average a molecule will have velocity c, say, and this will have components u, v and w in the direction of each of the axes respectively, as shown also.

From trigonometry,

$$c^2 = u^2 + v^2 + w^2$$

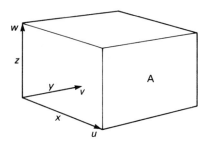

Figure 6.11

Consider next the *force* exerted on face A as a result of the *velocity component u*. The momenta before and after an impact with the wall of one molecule are mu and $-mu$ respectively.

Thus, the change in momentum is

$$\Delta p = mu - (-mu) = 2mu$$

Since the time interval Δt during which the molecule moves across the cube to face A and back is $\Delta t = 2L/u$, then *the change in momentum per second* is given by

$$\boxed{\frac{\Delta p}{\Delta t} = \frac{2mu}{2L/u} = \frac{mu^2}{L}}$$

Derivation of $P = \frac{1}{3}\rho\overline{c^2}$

From Newton's second law of motion, the force on face A due to one molecule is given by the rate of change of momentum of that molecule. Thus, the resulting pressure can be expressed as follows:

$$\boxed{\frac{\text{force on A}}{\text{area of A}} = \frac{mu^2}{L^3}}$$

But there are n molecules and they don't all happen to have the same velocity component u, so the *total pressure* on face A is

$$P = \frac{mu_1^2}{L^3} + \frac{mu_2^2}{L^3} + \ldots + \frac{mu_n^2}{L^3} = \frac{m}{L^3}(u_1^2 + u_2^2 + \ldots + u_n^2)$$

where u_1, u_2, etc. are all the different velocity components of the n molecules perpendicular to face A.

The *mean square velocity* is defined from the following expression:

$$\boxed{\overline{u^2} = \frac{u_1^2 + u_2^2 + \ldots + u_n^2}{n}}$$

As there is a cube of identical molecules, $\overline{u^2} = \overline{v^2} = \overline{w^2}$, and since from trigonometry $\overline{c^2} = \overline{u^2} + \overline{v^2} + \overline{w^2} = 3\overline{u^2}$ the following relationship is obtained:

$$\overline{u^2} = \tfrac{1}{3}\overline{c^2}$$

The total pressure on face A can now be rewritten in the simpler form

$$P = \frac{mn\overline{u^2}}{L^3} = \frac{mn\frac{1}{3}\overline{c^2}}{L^3}$$

The density of the gas is $\rho = \text{mass/volume} = nm/L^3$, therefore

$$P = \tfrac{1}{3}\rho\overline{c^2}$$

If the mass is constant and if the temperature is constant, then $\overline{c^2}$ will also be constant, i.e.

$$PV = nm\tfrac{1}{3}\overline{c^2} = constant \qquad (\text{since } V = L^3)$$

This then is the equation for *Boyle's law*. It also implies a link between the velocity or *speed* of molecules in a gas and the *temperature* of the gas.

Temperature and energy in an ideal gas

From the ideal gas equation,

$$PV = \frac{n}{N}\,RT$$

Then $\qquad \dfrac{n}{N}RT = \tfrac{1}{3}nm\overline{c^2}$

and $\qquad RT = \tfrac{1}{3}Nm\overline{c^2}$

Now this can be rewritten as

$$RT = \tfrac{2}{3}N \times \tfrac{1}{2}m\overline{c^2}$$

and $\qquad \tfrac{1}{2}m\overline{c^2} = \tfrac{3}{2}\dfrac{R}{N}T$

So the *average kinetic energy* of gas molecules in an ideal gas is directly proportional to the absolute temperature of the gas. Therefore as the *temperature rises* so does the kinetic energy of the molecules in the gas. The constant R/N is called **Boltzmann's constant**, k.

Hence the average kinetic energy is

$$E = \tfrac{3}{2}kT$$

The value for Boltzmann's constant can be obtained by

$$k = \frac{R}{N} = \frac{8.31}{6 \times 10^{23}} = \underline{1.38 \times 10^{-23}\,\text{J K}^{-1}}$$

A number of deductions can be made from this model of an ideal gas. However, only three will be discussed here.

Root mean square velocity (c_{rms})

The pressure and density of the ideal gas are related as follows:

$$P = \tfrac{1}{3}\rho\overline{c^2}$$

Therefore

$$\sqrt{\overline{c^2}} = \sqrt{\frac{3P}{\rho}}$$

Now for hydrogen gas at stp, $\rho = 0.09 \, \text{kg m}^{-3}$, and so

$$\sqrt{\overline{c^2}} = \sqrt{\frac{3 \times 0.76 \times 13\,600 \times 9.81}{0.09}} = \underline{1838 \, \text{m s}^{-1}}$$

This value is close to the speed of sound in hydrogen under the same conditions and indicates the motions of molecules in a gas are involved in the *propagation* of sound.

Avogadro's hypothesis

Equal volumes of gases under the same conditions of temperature and pressure contain equal numbers of molecules.

Consider two equal volumes V under the same pressure and temperature. Then

$$PV = \tfrac{1}{3} n_1 m_1 \overline{c_1^2} \quad \text{and} \quad PV = \tfrac{1}{3} n_2 m_2 \overline{c_2^2}$$

But since the *temperature* is the *same*, the *average molecular kinetic energies* of the two gases must be the *same*:

$$\tfrac{1}{2} m_1 \overline{c_1^2} = \tfrac{1}{2} m_2 \overline{c_2^2}$$

Therefore

$$\boxed{n_1 = n_2}$$

and *Avogadro's hypothesis is confirmed.*

Dalton's law of partial pressures

The pressure due to a mixture of gases or vapours which do not interact in any way is equal to the sum of their *partial pressures*.

Consider a volume V which contains three non-interacting gases:

$$P_1 V = \tfrac{1}{3} n_1 m_1 \overline{c_1^2} \qquad P_2 V = \tfrac{1}{3} n_2 m_2 \overline{c_2^2} \qquad P_3 V = \tfrac{1}{3} n_3 m_3 \overline{c_3^2}$$

As all gases in volume V are at the same temperature:

$$m_1 \overline{c_1^2} = m_2 \overline{c_2^2} = m_3 \overline{c_3^2} = m\overline{c^2} \quad \text{(say)}$$

By substitution and addition:

$$(P_1 + P_2 + P_3)V = \tfrac{1}{3}(n_1 + n_2 + n_3)m\overline{c^2}$$

But $(n_1 + n_2 + n_3)$ is the *total number* of molecules n and the total pressure P should be represented by

$$PV = \tfrac{1}{3} nm\overline{c^2}$$

Hence *Dalton's law.*

Formation of a vapour

Although the average speed of molecules in a liquid may not be high enough for most molecules to escape, a few of the fast ones do. This is *evaporation*. Since some of the fastest moving molecules have escaped, the *average* speed of the remaining molecules goes down. This means a drop in *temperature*, i.e. evaporation causes *cooling*.

For example milk coolers are soaked in water and as the water evaporates the milk is cooled. A sick person's brow is sprinkled with eau-de-cologne which also evaporates causing cooling. Water evaporates from the body as sweat and after exercise the body may overcool.

Factors affecting the rate of evaporation

(1)   An increase in temperature increases the rate of evaporation.
(2)   An increase in pressure above the liquid decreases the rate of evaporation.
(3)   An increase in the surface area of the liquid increases the rate of evaporation.
(4)   A draught removes molecules over the liquid surface, preventing their return to the liquid.

A liquid with a low boiling point which evaporates readily is said to be **volatile**.

Boiling and evaporation compared

Evaporation	Boiling
Surface effect	Takes place throughout the whole liquid
Occurs at all temperatures	Occurs only at the boiling point

Two types of vapour can be formed above the surface of a liquid: **saturated** and **unsaturated vapours**.

For the following examples assume $g = 9.8\,\mathrm{m\,s^{-2}}$.

Example 6.4

An ideal gas has volume $0.56\,\mathrm{m^3}$ when its pressure is 2.3×10^5 Pa. When the pressure is 6.7×10^5 Pa, what will be its volume assuming no change in temperature?

Applying Boyle's law,

$$PV = constant$$

Then $P_1 V_1 = P_2 V_2$

and so

$$V_2 = P_1 V_1 / P_2$$

Therefore

$$V_2 = \frac{2.3 \times 10^5 \times 0.56}{6.7 \times 10^5} = \underline{0.19\,\mathrm{m^3}}$$

Example 6.5

An ideal gas is kept at constant pressure as its volume is increased from $7.8\,\mathrm{m^3}$ to $11.3\,\mathrm{m^3}$ by heating. If the initial temperature of the gas is $67\,^\circ$C, what will its final temperature be?

Applying Charles' law,

$$V/T = constant$$

Then $V_1/T_1 = V_2/T_2$

and so

$$T_2 = V_2 T_1/V_1$$

Therefore

$$T_2 = \frac{11.3 \times (67 + 273)}{7.8} = \underline{493\,\text{K}}$$

(Note that thermodynamic temperatures must be used in gas laws calculations.)

Example 6.6

A gas has a density of $2.8\,\text{kg m}^{-3}$ at stp. If it is ideal, find the rms speed of the gas molecules at $135\,^\circ\text{C}$, if no change in volume occurs during heating.

Applying the pressure law,

$$P/T = constant$$

Then $P_1/T_1 = P_2/T_2$

and so

$$P_2 = P_1 T_2/T_1$$

Therefore

$$P_2 = \frac{1.01 \times 10^5 \times (135 + 273)}{273} = 1.51 \times 10^5\,\text{Pa}$$

From the kinetic theory,

$$P = \tfrac{1}{3}\rho \overline{c^2}$$

Thus,

$$\sqrt{\overline{c^2}} = \sqrt{\frac{3P}{\rho}} = \sqrt{\frac{3 \times 1.51 \times 10^5}{2.8}} = \underline{402\,\text{m s}^{-1}}$$

Example 6.7

Determine, for the gas in example 6.5, the ratio of the average molecular kinetic energy at stp to that at its new temperature, and hence find the rms speed of the gas molecules at stp.

The mean molecular KE is

$$\overline{E} = \tfrac{3}{2}kT = \tfrac{1}{2}m\overline{c^2}$$

Thus, the ratio of mean kinetic energies is given by

$$\overline{E}_1/\overline{E}_2 = T_1/T_2 = 273/408 = 0.67$$

Similarly, the ratio

$$\sqrt{\overline{c_1^2}}/\sqrt{\overline{c_2^2}} = \sqrt{T_1/T_2}$$

Therefore

$$\sqrt{\overline{c_1^2}} = \sqrt{\overline{c_2^2}} \times \sqrt{T_1/T_2} = 402 \times \sqrt{0.67}$$

Thus,

$$\sqrt{\overline{c_1^2}} = \underline{329\,\text{m s}^{-1}}$$

Example 6.8

Determine the volume of liquid X, which has density $1100\,\text{kg m}^{-3}$, that could be obtained by liquefying $0.1\,\text{m}^3$ of gas X. The gas has a pressure of 1.01×10^7 Pa and a temperature of 280 K. (Assume $R = 8.3\,\text{J mol}^{-1}\,\text{K}^{-1}$ and that the mass of 1 mol of X is 17 g.)

For an ideal gas,

$$PV = nRT$$

Therefore

$$n = PV/RT$$

Thus, the number of moles of X is

$$n = \frac{1.01 \times 10^7 \times 0.1}{8.3 \times 280}$$

So $\quad n = 435\,\text{mol}$

Thus, mass of X

$$m = 435 \times 0.017 = 7.4\,\text{kg}$$

Therefore volume of liquid

$$V = \frac{m}{\rho} = \frac{7.4}{1100} = \underline{6.7 \times 10^{-3}\,\text{m}^3}$$

Example 6.9

A diver's breathing apparatus consists of two cylinders of equal volumes V which are joined together by fine tubing that can be closed off. During tests the apparatus is kept at a temperature of 35 °C and one cylinder is evacuated. If the other contains gas, water vapour and liquid water at a total pressure of 9×10^4 Pa, find the pressure in the whole apparatus when the fine tubing is opened. (Assume water vapour pressure at 35 °C is 1×10^4 Pa.)

Applying Boyle's law,

$$PV = constant$$

Then $\quad P_1 V_1 = P_2 V_2$

and so

$$P_2 = P_1 V_1/V_2$$

Therefore

$$P_2 = \frac{(9-1) \times 10^4 \times V}{2V} = 4 \times 10^4\,\text{Pa}$$

(V has been taken as the volume of one cylinder.)

However, the vapour pressure due to the liquid water still remains. Thus, final pressure

$$P = (4 \times 10^4) + (1 \times 10^4) = \underline{5 \times 10^4\,\text{Pa}}$$

Example 6.10

How much work is done in pushing out the atmosphere when 3×10^{-2} kg of water are converted to steam at 373 K by the application of heat to the water? (Densities of water and steam at 373 K are $1 \times 10^3\,\text{kg m}^{-3}$ and $0.5\,\text{kg m}^{-3}$ respectively, at atmospheric pressure.)

External work done by gas:

$$W = (V_2 - V_1)P$$
$$V_1 = m/\rho_1 = 3 \times 10^{-2}/1 \times 10^3 = 3 \times 10^{-5}\,\text{m}^3$$
$$V_2 = m/\rho_2 = 3 \times 10^{-2}/0.5 = 6 \times 10^{-2}\,\text{m}^3$$

The pressure P is atmospheric pressure: $P = 1.01 \times 10^5$ Pa. Therefore

$$W = [(6 \times 10^{-2}) - (3 \times 10^{-5})] \times 1.01 \times 10^5$$

Thus,

$$W = \underline{6057\,\text{J}}$$

Saturated and unsaturated vapours

A **saturated** vapour is a vapour at a certain temperature which has a *maximum* number of molecules above the surface of the liquid, i.e. this is a vapour in *dynamic equilibrium* with its own liquid. (The number of molecules leaving the liquid per second in evaporation is the same as the number returning per second to the liquid.)

Thus the pressure of the vapour (called the **saturated vapour pressure, svp**) is constant for that temperature of container. If the temperature is increased more molecules leave and return to the liquid per second, so the pressure rises.

An **unsaturated vapour** is a vapour *not* in dynamic equilibrium with its liquid. It *obeys* the *gas laws approximately*, but not near saturation.

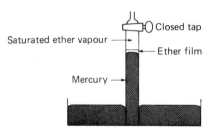

Figure 6.12

If ether is gradually introduced into a simple mercury *barometer* vacuum by means of a tap (Fig. 6.12), then the mercury level in the tube drops. This is because unsaturated vapour formed by the ether exerts a pressure on the mercury. If more ether is added until a liquid ether film is seen, the mercury level drops further and then stops. At this point the ether vapour is *saturated*. No more vapour can be made.

The total drop in the mercury level gives the svp at *that temperature*. If the vapour is warmed the mercury level begins to fall again, so

svp *increases* with *temperature*

Boiling occurs in a liquid when the *svp is equal to the external pressure.*

Also as temperature is increased the density of a saturated vapour increases while that of the liquid decreases. When the two densities are the same, the critical temperature has been obtained.

Behaviour of real gases

Real gases only obey the gas laws at very low pressures.

Andrews' experiment

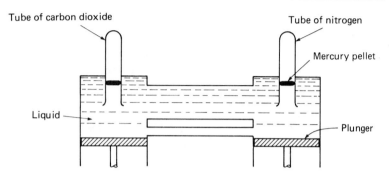

Figure 6.13

Carbon dioxide gas was contained in a calibrated glass tube by a mercury pellet. The open end of the tube was set in a chamber of liquid, to which *pressure* was applied by screwing in plungers. The pressure was measured from the *variation in volume* of air or nitrogen in a similar tube next to the one containing carbon dioxide. (Andrews assumed the nitrogen to *obey Boyle's law*.) The experiment was repeated for *different* temperatures of gas by surrounding the exposed part of the glass tube with a water bath (i.e. the tube with carbon dioxide). See Fig. 6.13.

The results of the experiment can be illustrated in the form of a graph of pressure versus carbon dioxide volume (Fig. 6.14). The curves that make up the graph are called *isothermals* since they record the variation of pressure with volume at one particular temperature.

Results of Andrews' experiment

Andrews was able to obtain the variation of pressure and volume for carbon dioxide as shown in Fig. 6.14.

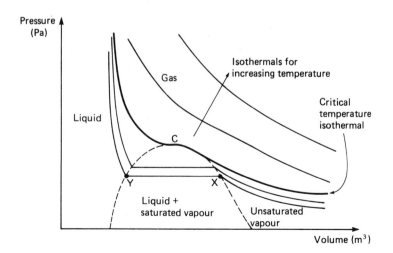

Figure 6.14

Notes from the experiment

(1)   As Andrews increased the pressure along the isothermal to point X on the diagram, a meniscus was formed above the mercury pellet in the carbon dioxide tube.

(2)   This indicated *liquid formation* but as he pushed the plungers in further along XY no change in pressure resulted, although volume decrease did occur (*saturation* of vapour at X).

(3)   However, the meniscus did rise and disappeared at the top of the tube when point Y was reached.

(4)   From then on the pressure increased rapidly, indicating that only liquid was left above the mercury.

(5)   Repeating the experiment along the *isothermal* to point C Andrews was unable to observe a meniscus. This was the isothermal for the *critical temperature*, at which the density of the vapour and liquid is the same.

(6)   Isothermals for higher temperatures were found to obey *Boyle's law* fairly well.

Critical temperature

The **critical temperature** is the temperature above which a gas cannot be liquefied by an increase in pressure.

Critical point

At the **critical point** there is a critical pressure and volume and the density of the liquid is equal to the vapour density.

Gas

A **gas** is a substance in a gaseous state above its critical temperature.

Vapour

A **vapour** is a substance in a gaseous state below its critical temperature.

Triple point

On a pressure versus temperature diagram, curves can be plotted showing the *equilibrium conditions* for the physical states of a substance: solid, liquid and gas or vapour (Fig. 6.15).

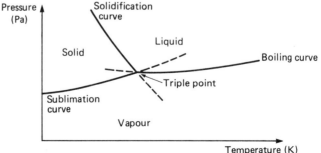

Figure 6.15

The **triple point** is where the three states or *phases* of the substance can *coexist*. The triple point for water is at 273.075 K.

Specific heat capacities of a gas

For a given amount of heat the temperature rise of a gas is determined by the external work it does, if any, in expansion. Thus a gas has an *infinite* number of heat capacities. The two *principal* ones are as follows.

Specific heat capacity under constant volume, c_v

The heat required to raise the temperature of 1 kg of gas by 1 K at constant volume.

Specific heat capacity under constant pressure, c_p

The heat required to raise the temperature of 1 kg of gas by 1 K at constant pressure.

Note: Both heat capacities can be stated for *one mole of gas* in which case they are called **molar heat capacities**.
Since work is done to push out the walls of a container,

$$c_p > c_v$$

For *one mole* of ideal gas $c_p - c_v = W$, where W is the work required to push out the walls of the container. This is *external work*. Now,

$$\text{work} = \text{force} \times \text{distance moved in direction of force}$$

$$= \frac{\text{force} \times \text{volume change}}{\text{area over which force is applied}}$$

If the volume changes from V_1 to V_2 this can now be written as an integral:

$$\text{work } W = \int_{V_1}^{V_2} P \, dV$$

Considering *constant pressure* and a temperature change of 1 K, then

$$PV_1 = RT \quad \text{and} \quad PV_2 = R(T+1)$$

and thus

$$W = P(V_2 - V_1) = R$$

Therefore

$$\boxed{c_p - c_v = R}$$

which is another definition of R *in terms of work*.

Ratio of specific heats for a gas (atomicity), γ

$$\boxed{\gamma = c_p / c_v}$$

For monatomic gases (e.g. neon), the atomicity is 1.67. For diatomic gases such as oxygen it is 1.40, while for polyatomic gases such as methane it lies between 1.00 and 1.40.

Molecular theory of matter

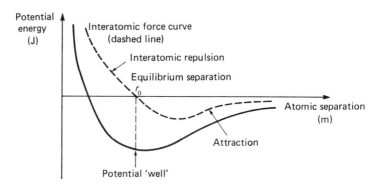

Figure 6.16

The *kinetic theory* proposes that matter consists of atoms or molecules which are always in a state of agitation. The energy which they have is called *internal energy*. This has two parts:

(1) *electrical potential* energy due to electrostatic forces between the atoms,
(2) *kinetic* energy which is dependent upon temperature.

Solids

In solids atoms have some sort of *ordered arrangement*, e.g. a crystal lattice. They vibrate about an equilibrium distance r_0 from each other, where the forces on them are zero. Their *internal* energy is increased when heat is transferred to them. This can have two effects: (a) a rise in *temperature*, and (b) an increase in *electrical potential* energy. The latter enables an atom to overcome some of the attractive forces upon it, and so the equilibrium distance between the atoms *increases*. (*Expansion* takes place.)

Internal energy in a solid is wholly *vibrational energy*.

If enough heat energy is supplied to the solid, the atoms begin to move *freely*, although they are still influenced by their neighbours. This is a *liquid* and the energy required to break up the ordered atomic arrangement is called the *latent heat of fusion*.

Specific latent heat of fusion

This is the heat required to convert 1 kg of solid into liquid under *constant temperature* (units $J\,kg^{-1}$).

The internal energy of a liquid is both *vibrational* and *translational* due to the movement of atoms. If even more heat is added, the atoms of the liquid begin to overcome the attractive forces between them. Eventually at the *boiling point* temperature they become completely independent of each other and can be large distances apart. This is a *gas*, and the atoms have high speeds but very small interatomic forces—zero potential energy.

Virtually all the internal energy of a gas is *kinetic*.

Specific latent heat of vaporisation

This is the heat required to convert 1 kg of liquid into vapour at *constant temperature*.

Apart from the transfer of heat, the internal energy of a body can be changed by *doing work*. For example compression of air in a bicycle pump leads to a rise in temperature, while rapid expansion as experienced by gas in the spray of an aerosol can lead to cooling.

Specific latent heat of vaporisation using Berthelot's method

A coil carrying current I and across which there is a potential difference V heats the liquid under investigation in a perforated bulb (see Fig. 6.17). Vapour escapes through the perforations and passes down through a *condenser*, which has a *cold water* inlet. As it does so, the vapour turns to liquid and is collected in a beaker whose mass is known. After time t the mass of liquid m collected is found.

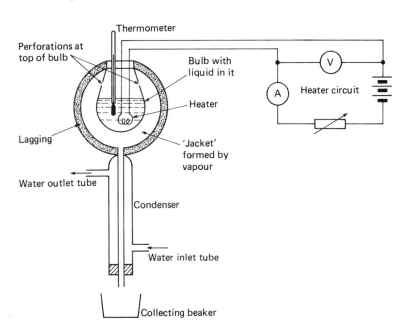

Figure 6.17

The 'jacket' of vapour produced in the apparatus *reduces heat losses* and occurs when the rate of vaporisation is the same as that of condensation, i.e. when a *steady state* has been achieved. Under this condition all the heat supplied is used as latent heat or to cover heat losses from the jacket itself. The experiment is repeated for the same length of time but using different values of current I' and potential difference V'. For the first and second experiments:

$$IVt = mL_v + H \quad \text{and} \quad I'V't = m'L_v + H$$

respectively, where L_v and H are the specific latent heat of vaporisation and heat loss respectively. Thus,

$$(IV - I'V')t = (m - m')L_v$$

Therefore the heat loss H has been *eliminated* and the specific latent heat of vaporisation is given by

$$L_v = \frac{(IV - I'V')t}{m - m'}$$

First law of thermodynamics

For an ideal gas the *internal energy* is entirely *kinetic* since for an ideal gas the intermolecular forces are negligible and therefore the potential energy is negligible.

But for real gases this is not so. When a small quantity of heat δQ is supplied to a real gas there is an increase in the kinetic energy of the molecules δE. Also internal work is done against molecular attractions, with a resulting gain in potential energy δI.

Therefore the total energy change for the *internal energy* is

$$\delta U = \delta E + \delta I$$

($\delta I = 0$ for an ideal gas—*Joule's law*.)

However, *external work* δW is done also by the gas as it pushes back a wall or piston and therefore, applying the *conservation of energy principle*,

$$\delta Q = \delta W + \delta U$$

This is the **first law of thermodynamics**. It can be stated a number of ways and is founded upon the idea of conservation of energy. One statement of it is as follows.

If the energy of a closed system is changed by an amount δQ, that change is equal to the sum of the changes in internal energy and the external work done.

Q is positive or negative if the heat is supplied or removed respectively. Similarly δW is positive or negative if the external work done by the gas is by *expansion* or by *compression* respectively.

Four different processes can now be defined for pressure–volume variation in a gas.

(1) **Isovolumetric** (constant volume) $\delta V = 0$ so $\delta W = 0$ and

$$\delta U = mc_v(T_2 - T_1)$$

(2) **Isobaric** (constant pressure) $\delta Q = \delta U + \delta W$. Therefore

$$mc_p(T_2 - T_1) = mc_v(T_2 - T_1) + P(V_2 - V_1)$$
$$\text{from (1)}$$

(3) **Isothermal** (constant temperature) $\delta U = 0$ so $\delta Q = \delta W$ and

$$PV = constant$$

(4) **Adiabatic** (no change in heat energy) $\delta Q = 0$ so $\delta W = -\delta U$.

Reversible change

This can be isothermal or adiabatic. It is an ideal change carried out infinitely slowly so that the system is in equilibrium at any instant.

Pressure–volume graphs for gases

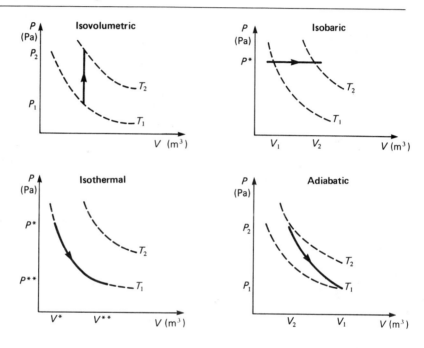

Figure 6.18

Temperature

This is the property which indicates how well matter is able to transfer energy by conduction or by radiation.

A thermometer measures temperature. It does this by measuring a property of matter which varies with temperature; such a property is called **thermometric**. Some of these properties are shown below with their corresponding thermometers.

Thermometric property	Example of thermometer
Volume expansion	Mercury-in-glass
Pressure exerted by gas	Constant volume gas
Electrical resistance	Platinum resistance
Vapour pressure	—
Thermoelectric effects	Thermocouple
Emission of radiation	Pyrometer
Magnetism	—

Temperature scales

A temperature scale is based on a *particular thermometric property*. It is defined by means of *two fixed points*, and the interval between these, called the *fundamental interval*, defines temperature on the scale.

Fixed points

An *equilibrium temperature* which can be *accurately reproduced* anywhere is called a *fixed point*. For example the temperature at which pure boiling water is in equilibrium with the steam above it when atmospheric pressure is 0.76 m Hg.

Ideal gas scale

The two fixed points here are *absolute zero* and the *triple point* of water, 273.16 K. The scale depends upon the properties of gases in general. It is determined by use of the *constant volume gas thermometer* and the unit of temperature on it is the *kelvin*.

Celsius scale

The two fixed points are the *ice point* where pure ice melts and the *steam point* at which pure boiling water is in equilibrium with the steam above it at a pressure of 0.76 m Hg. The division of the fundamental interval into one hundred equal parts was first proposed by *Celsius* and each unit was called a *degree celsius* or a *degree centigrade* (°C).

Absolute thermodynamic scale

This is a *theoretical* scale and has the advantage over others of being independent of the *property of any substance*. It defines the SI unit of temperature, the *kelvin*.

International practical temperature scale (IPTS)

This is based on *eleven primary fixed points*, and thermometers using different thermometric properties over *certain* temperature *ranges* on it employing internationally agreed *formulae*. It is more practical to measure widely varying temperatures on this scale than on the ideal gas scale. Both scales *must agree* at the *fixed points*, and differences elsewhere are small.

Fixed point	°C	K
Steam point	100.00	373.15
Triple point of water	0.01	273.16
Ice point	0.00	273.15
Absolute zero	−273.15	0.00

Thermometric property

Consider P_0 and P_{100} as the values of a *certain* property measured by a thermometer at the *ice* and *steam points*. If this thermometer measures a value P_θ then the temperature in °C is given by

$$\theta = \left(\frac{P_\theta - P_0}{P_{100} - P_0}\right) \times 100$$

There is no reason why P_θ should be the same for different types of thermometer and so a given temperature may be given different values if it is measured by different thermometers.

All scales of temperature *must agree* at the *ice* and *steam points*.

Liquid-in-glass thermometer

The formula used in this case is

$$\theta = \left(\frac{L_\theta - L_0}{L_{100} - L_0}\right) \times 100$$

where L represents the length of the *liquid column*. It has a range of 230 K to 570 K with mercury, and *expansion* of the liquid is the thermometric property used. The device consists of a bulb containing the liquid and a graduated capillary of glass through which the liquid rises when it expands out of the bulb.

Clinical thermometers use alcohol and have a kink in the thread to delay the liquid from returning to the bulb after the temperature has been taken.

Advantages of mercury
Even expansion; clearly seen; mercury is a good conductor of heat.

Disadvantages of the thermometer
Bore is not uniform; exposed stem when measuring; glass envelope expands; 'creep' of zero due to bulb expansion.

Simple constant volume gas thermometer

The thermometer is set up as shown in Fig. 6.19 in the simple version. A *water bath* surrounds the bulb containing the *gas*. Uniform heating of the

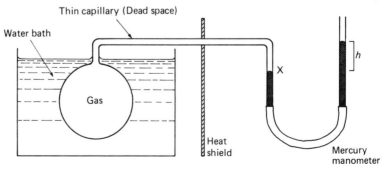

Figure 6.19

gas is achieved by *stirring* the water when it has reached the required temperature. In the school laboratory *dry air* is used in the bulb, but for accurate measurements *nitrogen* or *hydrogen* is more suitable.

A fine glass *capillary tube* connects the bulb to a moveable mercury reservoir via rubber tubing. At different temperatures, when measuring the pressure from this mercury manometer, the height of the reservoir is adjusted so that X *remains the same* all the time. (The tubing may be fitted with some sort of level *index* at X to make this easier.)

The pressure recorded is $P = H + h$, where H is *atmospheric pressure*. Again the conversion to °C is achieved by the use of the formula

$$\theta = \left(\frac{P_\theta - P_0}{P_{100} - P_0} \right) \times 100$$

The *ice point* and *steam point* pressures are established by packing pure melting ice around the bulb instead of the water bath, and by heating the water in the water bath until it boils, respectively.

Advantages
Wide range—70 K to 1770 K for nitrogen gas; very accurate.

Disadvantages
Cumbersome; slow to use since the gas pressure takes time in building up (time lag); bulb has large heat capacity.

Thermocouple

When a closed circuit made of two dissimilar metals or alloys has the two junctions at different temperatures a current flows in the circuit. This is the **Seebeck effect** and is the basis of the **thermocouple**.

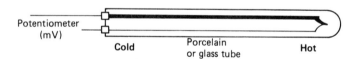

Figure 6.20

In the school laboratory a *potentiometer* is used to measure the *thermoelectric emf* produced across the thermocouple (Fig. 6.20). The *cold junction* is usually kept in ice, but for measuring high temperatures it is not necessary to do that. Thermoelectric emf does not vary *linearly* with temperature and the formula used is

$$E = a + b\theta + c\theta^2$$

where *a*, *b* and *c* are constants.

Thermoelectric emf and the thermocouple

If the thermoelectric emf produced by a thermocouple is plotted against temperature, the characteristic curve in Fig. 6.21, called a *parabola*, is obtained.

Thermocouples are not used beyond their *neutral temperatures* as a thermoelectric emf E^* would indicate *two different* temperatures.

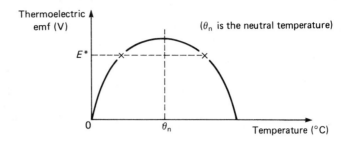

Figure 6.21

The *Seebeck effect*, which is the underlying explanation for the thermoelectric emf, arises because the *electron density* is not the same for different metals and depends on temperature. So electron 'diffusion' at the junctions of the thermocouple occurs at *different rates*. Hence the net effect is a *current and an emf*.

An arrangement of many thermocouples is called a *thermopile*.

Advantages
Wide range; small heat capacity (no time lag in measurements); most accurate thermometer in the range 900 K to 1340 K; can measure temperature over a small area.

Platinum resistance thermometer

The *resistance* of pure metals *increases* as the temperature *rises*. This thermometer consists of platinum wire wound onto a mica strip placed inside a porcelain container. To overcome any resistance arising from the heating of leads connected to the thermometer, a pair of *dummy leads* forms part of the circuit making up the device. These are heated along with the thermometer. The resistance is measured on a *Wheatstone-type bridge* circuit (Fig. 6.22).

The formula used is

$$\theta = \left(\frac{R_\theta - R_0}{R_{100} - R_0} \right) \times 100$$

Advantages
Wide range—70 K to 1270 K; very accurate; best used for small differences of temperature, over area being investigated.

Disadvantage
Time lag due to high heat capacity of porcelain.

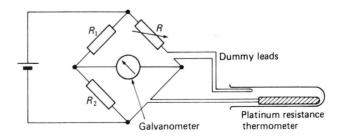

Figure 6.22

From circuit theory, $R_\theta = RR_2/R_1$, where the resistances are as shown in Fig. 6.22.

Beyond the range of platinum resistance thermometers several types of *pyrometer* are used.

Example 6.11

A hot liquid is poured into two identical glass dishes. (i) State the factors which determine the rate at which the temperature of the liquid falls and (ii) explain why, if twice as much liquid is poured into one dish as the other, the smaller amount of liquid will cool more quickly.

(i)   The factors which determine the rate of fall of temperature of the liquid are (a) the liquid temperature, (b) the liquid mass, (c) the thickness of the glass dish and (d) the temperature of the surroundings, and convection currents in the air.

(ii)   The rate at which both the amounts of liquid lose heat will be the *same*, although the larger body of liquid has the higher heat capacity. Thus, to cool to the *same extent* as the smaller amount of liquid, it needs to get rid of more heat, so it cools slowly.

Example 6.12

The platinum in a resistance thermometer has resistances of $5.8\ \Omega$ and $6.9\ \Omega$ at the ice and steam points respectively. When the thermometer indicates a temperature of $550\,°C$ determine the resistance of the platinum in it.

The temperature θ in $°C$ on the platinum resistance scale is expressed in terms of the resistance at that temperature R_θ as follows:

$$\theta = \frac{(R_\theta - R_0) \times 100}{R_{100} - R_0}$$

where R_{100} and R_0 are the resistances of the platinum at the steam and ice points respectively. Therefore

$$R_\theta = \frac{\theta(R_{100} - R_0)}{100} + R_0$$

Thus,

$$R_\theta = \frac{550 \times (6.9 - 5.8)}{100} + 5.8 = \underline{11.9\ \Omega}$$

Example 6.13

Discuss the characteristics which make a thermometric property a good property for measuring temperature.

For practical thermometers, the thermometric property must change in a regular, continuous way with temperature. The change must be large enough to enable fairly accurate measurements to be made. The property must also be unaffected by other physical conditions which are likewise dependent upon temperature.

Example 6.14

The length of the liquid thread in a liquid-in-glass thermometer is 11.7 cm at the steam point and 3.4 cm at the ice point. The length at temperature θ on a standard scale is given by

$$L_\theta = L_0(1 + 10^{-3}M\theta^2) \qquad (M = \text{constant})$$

Find the thread length when the thermometer is at $45\,°C$ on this scale.

Now, since
$$L_{100} = 0.034 \, (1 + 10^{-3} \, M \cdot 100^2)$$
then
$$M = 0.244 \, °C^{-2}$$
Therefore
$$L_{45} = 0.034 + 0.034 \, (0.244 \times 10^{-3} \times 45^2) = \underline{0.051 \text{ m}}$$

Disagreement of temperature scales

Different values of the same temperature are obtained from thermometers using different thermometric properties except at the fixed points. This is because, as the temperature changes, these properties vary differently. For example the pressure recorded by a constant volume gas thermometer when a mercury-in-glass thermometer reads 50 °C *is not* halfway between the pressure values at 0 °C and 100 °C. The only scale which is *independent* of a thermometric property is the absolute thermodynamic scale.

Mechanism of thermal conduction

Conduction is a transfer of heat in an unequally heated body without visible motion of any part of it. In a metal there are considerable numbers of *conduction* electrons per unit volume. These are able to wander freely around the metal. If one part of the metal is heated, the conduction electrons there gain speed and an electron flow results. It causes a movement of charge and a *transfer of heat*. This cannot occur in an *insulator* since these have *few if any* conduction electrons. In both metals and non-metals heat is transferred also by the collisions of vibrating atoms. Since atoms are more massive than electrons, this process is much slower.

Convection

This is a transfer of heat in a fluid by the movement of the fluid itself. A heated area expands, becoming less dense and begins to rise. As it does so, it cools and sinks back into the body of the fluid. Thus, a *convection current* is produced.

Coefficient of thermal conductivity, *k*

A thin parallel sided lamina of thickness δx and area A has a temperature difference $\delta \theta$ between its two sides (Fig. 6.23).

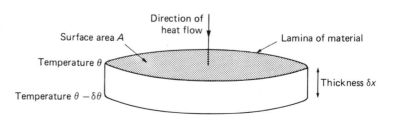

Figure 6.23

In the steady state heat flows normal to the parallel sides towards the lower temperature. The rate of flow increases with area A and temperature difference $\delta\theta$. It is inversely proportional to the thickness δx. Combining these relationships the rate of flow of heat is given by

$$\frac{\delta Q}{\delta t} = -kA\frac{\delta\theta}{\delta x} \quad \text{or} \quad \boxed{\frac{dQ}{dt} = -kA\frac{d\theta}{dx}}$$

where the thickness x and time t are made infinitesimally small. A minus sign is included to indicate the flow of heat to the lower temperature side.

k is the **coefficient of thermal conductivity** and is a constant for a particular material.

Thermal conductivity

$$\frac{dQ}{dt} = -kA\frac{d\theta}{dx}$$

From the above equation, the coefficient of thermal conductivity can be defined as the rate of flow of heat through a material per unit area per unit temperature gradient. (The units of k are $\mathrm{W\,m^{-1}\,K^{-1}}$.) $d\theta/dx$ is called the **temperature gradient**. In a steady state (temperatures not changing), a bar which is heated at one end and lagged (i.e. wrapped in a good insulator) has a *constant* temperature gradient (Fig. 6.24).

If the bar is *not lagged* there are heat losses through the sides and the temperature gradient is not constant (Fig. 6.25).

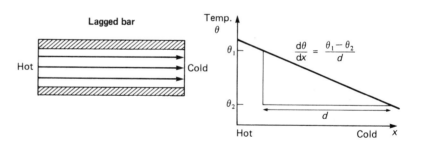

Figure 6.24

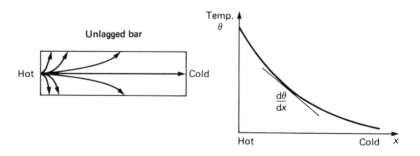

Figure 6.25

Since for the lagged bar the temperature gradient is constant, a graph of temperature θ versus distance along the bar will produce a straight line.

Thermal conductivity for a good conductor—Searle's method

The apparatus consists of a thick metal bar, one end of which is heated by a heating coil or steam chest. The bar itself is lagged and has holes for two thermometers so that the *temperature gradient* can be found. A coiled tube at the cooler end of the bar carries water from which the rate of flow of heat can be determined. Each end of this tube is fitted with a thermometer so that inlet and outlet water temperatures can be measured. See Fig. 6.26.

At the *steady state* when all the temperatures are constant, the mass of water flowing through the apparatus per second is found (M/t). Then

$$\frac{Q}{t} = \frac{M}{t} c_w (\theta_o - \theta_i)$$

where c_w is the specific heat capacity of water and θ_o and θ_i are the outlet and inlet water temperatures respectively.

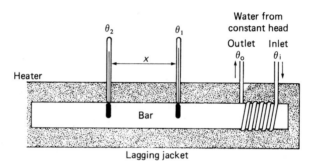

Figure 6.26

Heat is supplied to a lagged metal bar, from one end by means of a heating coil, and removed by water passing through tubing wrapped around the other end (Fig. 6.26). The thermometers in the bar are kept in mercury filled holes, to provide good thermal contact. The average diameter of the bar is determined by making several readings along the bar with calipers, from which the cross-sectional area A is obtained. At steady state, the thermometers are read and the mass of water passing through the tubing per unit time, M/t, is determined. (The average temperatures can be found by swaping over the thermometers, and the distance between them, x, is also found. Since

$$Q/t = Mc_w (\theta_o - \theta_i)/t$$

where c_w is the specific heat capacity of water, therefore as

$$Q/t = kA(\theta_2 - \theta_1)/x$$

then

$$k = \frac{Mc_w (\theta_o - \theta_i)}{tA (\theta_2 - \theta_1)} x$$

Thermal conductivity for a bad conductor—Lee's disc

The apparatus consists of a steam chest with a thick brass bottom placed on top of a circular wafer of the material under investigation which itself

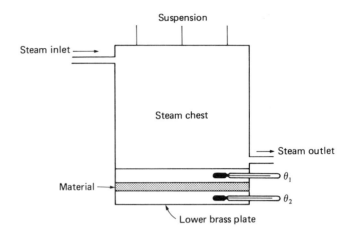

Figure 6.27

rests on a lower brass plate (Fig. 6.27). The whole apparatus is suspended from a stand. Each of the brass plates has a thermometer and the average thickness of the material x is found by making readings using a micrometer at several points on it.

Steam is allowed to pass into the steam chest and the temperatures from the thermometers are noted when the steady state has been achieved. The steam chest and the material are next removed and the lower plate heated using a bunsen until its temperature is a little above its previous steady state value. The bunsen is then removed and a felt pad placed on top of the plate which is allowed to cool.

As it cools its temperature is monitored at regular intervals of time. The readings are then used to plot a cooling curve, from which the slope S at the steady state temperature of the lower plate is determined. Thus, if A is the cross-sectional area of the material wafer,

$$Q/t = kA(\theta_1 - \theta_2)/x = m_p c_p S$$

Hence

$$\boxed{k = \frac{m_p c_p S x}{kA(\theta_1 - \theta_2)}}$$

Example 6.14

A sheet of balsa wood has area $5\,\text{m}^2$ and is 4 mm thick. If the temperature difference across its two largest surfaces is $21°C$ and the thermal conductivity of balsa is $0.07\,\text{W}\,\text{m}^{-1}\,\text{K}^{-1}$, find the rate of transfer of heat through the balsa.

Applying the steady state equation,

$$\frac{Q}{t} = -kA\frac{\Delta\theta}{x} = -\frac{0.07 \times 5 \times 21}{4 \times 10^{-3}} = \underline{-1.84\,\text{kJ}\,\text{s}^{-1}}$$

Example 6.15

The lagging on a large industrial boiler consists of two rubber-like sheets each of thickness 3 mm separated by polystyrene 12 mm thick. The ratio of the thermal conductivity of the rubber to that of the polystyrene is $40:1$. Compare (i) the temperature gradient in the rubber to that in the polystyrene and (ii) the temperature differences across one sheet of rubber and the polystyrene.

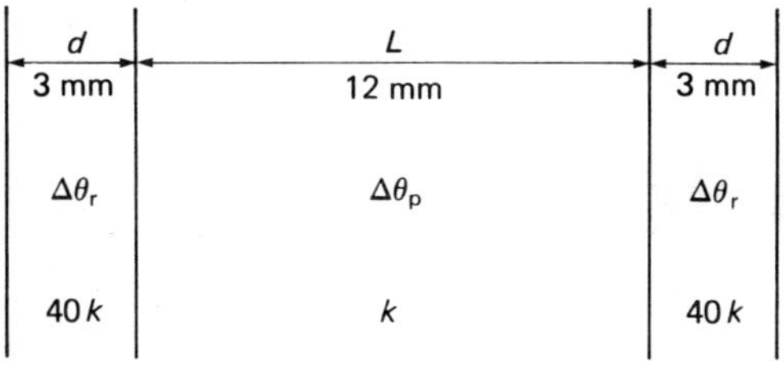

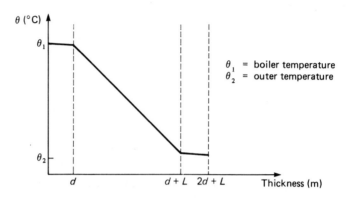

Figure 6.28

Figure 6.29

(i) See Figs 6.28 and 6.29. Let k and A represent the thermal conductivity of the polystyrene and the surface area of one rubber sheet respectively.
   Applying the law of conservation of energy,

$$\frac{dQ}{dt} = \frac{-40kA\,\Delta\theta_r}{d} = \frac{-kA\,\Delta\theta_p}{L} = \frac{-40kA\,\Delta\theta_r}{d}$$

Considering the last two parts of the above equation, then

$$\frac{\Delta\theta_r/d}{\Delta\theta_p/L} = \frac{1}{40} = \underline{0.025}$$

(ii) From the answer to part (i), then

$$\Delta\theta_r/\Delta\theta_p = 0.025d/L$$

Therefore

$$\Delta\theta_r/\Delta\theta_p = \underline{6.25 \times 10^{-3}}$$

7
Electricity

Introduction

This chapter covers electrostatics and direct current. It starts by stating the basic law of electrostatics and defining the main types of material used in its study, i.e. conductors and insulators. The use of an electroscope and charging by induction follow. The discussion of the electrophorus leads to the idea of electric potential energy and charge distribution. The diagrammatic representation of electric fields and Faraday's ice pail experiment are followed by a treatment of electric potential and potential difference. The next section deals with capacitance and practical capacitors. A mathematical look at electrostatics concerning Coulomb's law, field strength and potential gradient with a consideration of the equivalence of electrostatics and current electricity conclude this topic.

The last part of the chapter gives basic definitions and expressions for current electricity. The following sections deal with Ohm's law, resistor networks and resistivity. The different types of resistor and the effect of temperature are also described. The chapter is concluded with a section on one of the most important instruments used in basic electricity—the potentiometer and a brief look at electrolysis.

Revision targets

The student should be able to:

(1) Describe the basic phenomena and facts of electrostatics.
(2) Explain what is meant by capacitance and describe the factors which affect the capacitance of a parallel plate capacitor.
(3) State and apply Coulomb's law to numerical problems.
(4) Define field strength and electric potential.
(5) Apply the ideas of field strength and potential quantitatively to problems.
(6) Explain what is meant by potential gradient.
(7) Discuss the relationship between electrostatics and current electricity.
(8) Define drift velocity and current.
(9) Apply the basic expressions for current to numerical problems.
(10) Define and distinguish the terms potential difference and electromotive force.
(11) State Ohm's law and discuss the exceptions to it.
(12) Calculate for current, p.d., resistance and emf in problems dealing with resistor networks and simple circuits.

(13)   Discuss qualitatively and quantitatively the factors which determine the resistance of a conductor.

(14)   Explain what precautions are taken to make sure a potentiometer has been set up correctly and how to calibrate it.

(15)   Describe the advantages a potentiometer has over conventional meters and how it can be used to compare emf's.

(16)   Describe how the potentiometer can be used to calibrate a voltmeter and an ammeter.

(17)   Explain how the potentiometer is calibrated and used for the determination of thermoelectric emf's.

(18)   Explain what is meant by the internal resistance of a cell and how it can be determined using a potentiometer.

(19)   With reference to Ohm's law explain what is meant by back emf.

(20)   Define the term charge to mass ratio.

Electrostatics

When two materials which have been rubbed together are attracted to each other, **electrostatic** attraction is occurring. The materials have become *charged* with static electricity. There are two types of electrostatic charge—**positive** and **negative**. Conventionally, anything which repels *ebonite* which has been stroked with fur is *negatively* charged. Repulsion of *glass* stroked with silk indicates a *positive* charge.

Polythene and cellulose acetate are sometimes used instead of ebonite and glass respectively.

Material		Rubbed with	Charge acquired
Polythene	(white rod)	Fur	Negative
Ebonite	(black rod)	Fur	Negative
Glass		Silk	Positive
Cellulose acetate	(clear strip) of plastic)	Silk	Positive

By freely suspending a rod of the material that has been charged and bringing up to it another which has also been charged, the following law of force between electric charges can be deduced.

> Like charges repel; unlike charges attract.

Basically an atom is made up of a small positive *nucleus* and around this is a 'cloud' of *negatively* charged electrons. For the atom to be *neutral*,

> number of positive charges = number of electrons

A *transfer of electrons* occurs when bodies are charged by rubbing. If the body *gains* electrons it becomes *negatively* charged and if it *loses* electrons it becomes *positively* charged. When two bodies are charged by rubbing them together,

negative charge on one body = positive charge on other

Conductor

A **conductor** has electrons or other charged particles which can move with ease throughout that body.

Insulator

An **insulator** is a body in which electrons and other charged particles are not able to move about freely.

Electroscope

The gold leaf electroscope detects and measures small electric charges (Fig. 7.1).

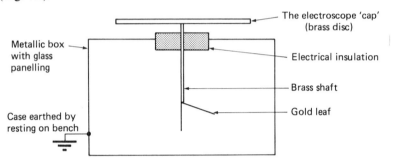

Figure 7.1

Experiments on the electroscope

(1)   When a charged body is brought near to the electroscope, but not touching it, a *temporary* charge is *induced* onto the electroscope cap. This charge disappears when the charged body is removed.

(2)   Charging the electroscope cap by actually touching it with a charged body, i.e. by *contact*, often gives the opposite charge required.

(3)

Charge on electroscope	Charge brought near to cap	Charge in leaf separation
Positive	Positive	Gets larger
Negative	Negative	Gets larger
Positive	Negative	Gets smaller
Negative	Positive	Gets smaller
Either	Neutral	Gets smaller

It can be seen that only *an increase* in the leaf separation (or divergence) indicates the type of charge applied.

(4)   When different materials are placed onto a charged electroscope with the hand, the following deductions can be made.

(i)   **Good insulator**—no leakage of charge to Earth through the hand and thus the leaf divergence does not change.

(ii)   **Good conductor**—instant collapse of leaf as all the charge passes to Earth via the hand.

(iii)   **Poor conductor/poor insulator**—a gradual collapse of the leaf.

Electrostatic induction

This occurs when a charged body is brought near to a neutral or uncharged body. The induced charge is always *opposite* in sign to that of the *inducing* charge used.

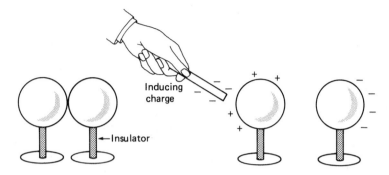

Figure 7.2

For example, if a negatively charged rod is brought up to two insulated metallic spheres which are in contact but neutral, a movement of electrons occurs. (See Fig. 7.2.) When the spheres are separated, still holding the rod nearby, the charges on the respective spheres are trapped. So now if the rod is moved away each sphere retains its charge.

When a transfer or movement of charge occurs in electrostatics the electron is always responsible; nuclear positive charge is too tightly bound.

Proof plane

This is made up of a small metal disc and an insulating handle (Fig. 7.3). It is used for transferring charge. Charge is transferred to an electroscope by bringing the charged proof plane down slowly, from a great height, onto the cap. Both the electroscope and the proof plane are 'discharged', i.e. lose their charge, by touching them.

Figure 7.3

Charging by induction

An almost unlimited number of charges from a single inducing charge can be created without any loss of inducing charge. Electrons are repelled by bringing up a negatively charged rod to a conductor (Fig. 7.4). If the rod is kept near and the opposite end of the conductor earthed by *touching it*, electrons are repelled to Earth, leaving a positively charged conductor. (The charged rod is always held near to the conductor until the fingers earthing the conductor have been removed.)

A negatively charged conductor can be obtained by applying a positively charged rod in a similar way.

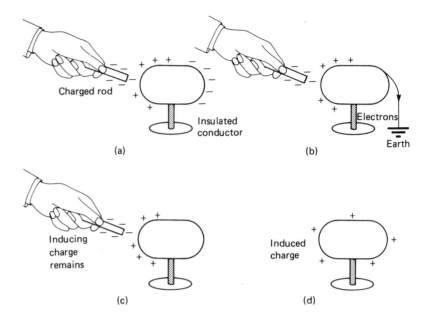

Figure 7.4

Electrophorus

This instrument works by electrostatic induction and can be used to produce many positive charges from one negative charge. It is made up of a slab of insulating material, usually polythene, and a proof plane (Fig. 7.5).

The slab is given a negative charge by stroking it with fur. The proof plane is then placed on top of this, and electrostatic induction occurs.

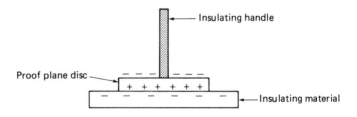

Figure 7.5

The proof plane disc is not quite flat and therefore does not become negatively charged by conduction, i.e. there are only small areas where it actually makes contact with the slab. While the disc is on the polythene it is earthed by touching it. When the disc is removed, it has a positive charge and sparking may occur between it and the polythene.

Electrical energy obtained from an electrophorus

The disc may be charged many times before the slab needs recharging. Electrical energy is only released when the disc is pulled away, since when it lies on the slab the total charge is zero. In pulling, *work* is done against the force of attraction between the disc and slab charges. It is this work that is converted to electrical energy.

Electric potential energy

The positive charge on an electrophorus disc gains electric potential energy, in a similar way that a stone gains gravitational potential energy as it is raised.

Charging an electroscope by induction

A charged rod is brought near to the electroscope cap so that the leaf diverges. Keeping the rod near, the electroscope cap is earthed by touching it and then the rod is taken away. This leaves the electroscope charged. See Fig. 7.6.

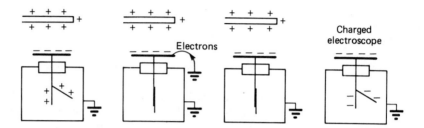

Figure 7.6

Electroscope precautions

If a strong negative charge is positioned well above a positively charged electroscope, electrons are repelled to the leaf and neutralise the positive charge there. The leaf collapses as the strong negative charge is lowered even further. At this point:

$$\frac{\text{number of free electrons}}{\text{at the leaf}} = \frac{\text{number of positive charges}}{\text{at the leaf}}$$

As the lowering continues more electrons are pushed onto the leaf, causing a divergence. Thus, all charged bodies must be brought down onto charged electroscopes from a good height. Otherwise, the collapse discussed above which occurs when the charged body and electroscope have different charges will not be seen. Only an increase in divergence will be observed, *wrongly* indicating that the charges are the same.

Distribution of charge

By inducing charge onto differently shaped conductors and employing a proof plane it can be shown that charge mainly concentrates at places which are *sharply curved* (Fig. 7.7).

Charge only resides on the *outside surface* of a hollow conductor.

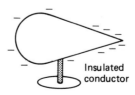

Insulated conductor

Figure 7.7

Faraday's butterfly net

The net is given a charge by induction. It is found that a proof plane will only transfer charge from the outside of the net, even when the net is pulled inside-out using the silk thread pull through (Fig. 7.8).

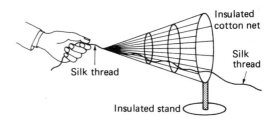

Figure 7.8

Discharging a rod

This can be done by quickly passing the rod over a flame. In the flame some of the gas molecules in the air lose some of their electrons and become **positive ions**. The free electrons can attach themselves to neutral molecules and these become **negative ions**. A charged rod attracts ions of opposite charge and is thus **neutralised** or **discharged**.

The air itself contains ions which have been produced by high energy radiation passing through the atmosphere, or from radioactive substances in the ground. Thus, even an insulated conductor loses its charge in time. It attracts ions of opposite charge and gradually becomes neutralised.

Action of highly charged points

A flame put near the end of a wire connected to a Van de Graaff generator is blown away from the wire. If the wire is put very near to the flame, the flame stretches towards and away from the wire (Fig. 7.9).

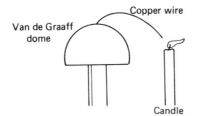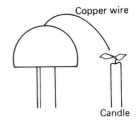

Figure 7.9

Since the air is partly ionised, ions will be attracted to the positively charged wire end. Positive ions will be repelled and will in turn 'knock' off electrons from other molecules to give even more positive ions. It is the motion or streaming of these which causes the flame to be *blown away*.

If the wire is put very close to the flame, negative ions in the flame itself are attracted to the wire while positive ones again are repelled. This 'stretches' out the flame.

Points also collect charge and an example is shown in Fig. 7.10 where an electroscope is charged.

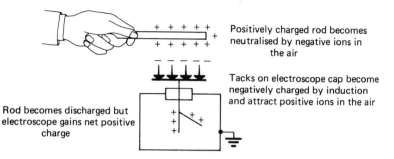

Positively charged rod becomes neutralised by negative ions in the air

Tacks on electroscope cap become negatively charged by induction and attract positive ions in the air

Rod becomes discharged but electroscope gains net positive charge

Figure 7.10

Lightning conductors

These use the action of points to protect tall buildings from lightning. The conductor is a thick copper strip on the outside of the building connecting metallic spikes that stick up at the top to a metal plate in the ground.

A negatively charged thundercloud repels electrons from the spikes leaving them positively charged and the metal plate negative. The plate being earthed loses its charge while an electric wind of positive ions produced by the spikes cancels some of the charge on the cloud. Any chance of lightning is reduced, and if it does occur it passes safely down the conductor to Earth.

Van de Graaff generator

This produces a continuous supply of positive charge on a large metal dome when a rubber belt is driven over a perspex roller.

Electric field

This is a region of force between charges and is represented by lines of force. The direction of the field at a point is the direction of the force on a *positive charge*. These lines start on a positive charge and end on an equal but negative one. (In diagrams both charges are not always shown.) The lines of force repel each other sideways and always try to shorten. See Figs 7.11 and 7.12.

Figure 7.11

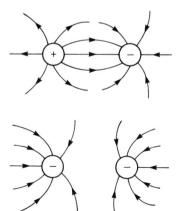

Figure 7.12

Faraday's ice pail experiment

A charged metal sphere on a long insulating handle is slowly lowered into a tall can resting on an electroscope without touching it. The leaf diverges and continues to do so if the sphere touches the bottom of the can. On removal, the sphere is found to have lost all of its charge.

Since all the field lines from the charges on the sphere end on the inside of the can they induce an equal charge of opposite sign on the inside. An equal charge of the same sign as that on the sphere is produced on the outside of the can. When the sphere touches the bottom of the can, the *inner* charge is *neutralised*. See Fig. 7.13.

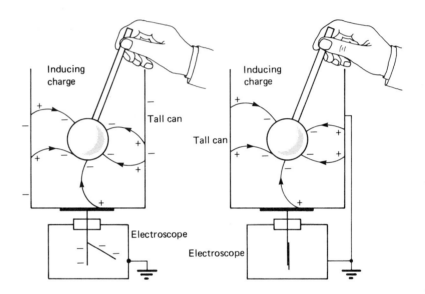

Figure 7.13

If the sphere is not allowed to make contact at all, but the can and electroscope are earthed by the application of a finger, the leaf collapses. Now, when the sphere is removed, the induced inner charge moves to the outside of the can and the leaf diverges by the same amount as it did before. Thus,

$$\frac{\text{magnitude of induced}}{\text{charge}} = \frac{\text{magnitude of inducing}}{\text{charge}}$$

Potential difference (p.d.)

If charge moves from one conductor to another when they are connected this is due to a potential difference between them.

In Fig. 7.14, electrons pass from Earth to the positively charged conductor because it is at a *higher* potential. Earth is taken as *zero potential*. For the negatively charged conductor, electrons flow to Earth.

Electrons flow from a lower potential to a higher potential.

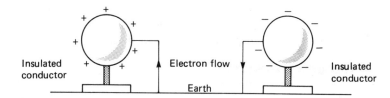

Figure 7.14

Presence of other charges

A charged conductor has a potential due to its own charge *and* due to other charges which may be nearby. For example a positively charged conductor can have a negative potential if there is a large enough negative charge near to it.

Potential

If two charges of the same sign are brought together, work is done and energy converted to electric potential energy.

Volt

This is the unit of p.d. About 500 V are needed for the divergence of an electroscope leaf.

Electroscope and p.d.

The divergence of the electroscope leaf *indicates a p.d.* between the leaf and the case of the electroscope, which is normally *earthed*. If a positive charge is brought near to an electroscope, it raises the potential of the leaf. The potential of the case is zero, and so the leaf diverges. Two other examples are shown in Figs 7.15 and 7.16.

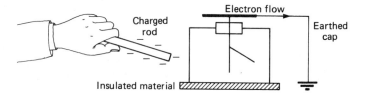

Figure 7.15

In Fig. 7.15, the leaf is at zero potential. The case is insulated and its potential is lowered by the presence of a negative charge. Since there is a p.d., the leaf diverges.

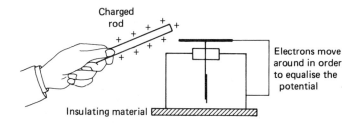

Figure 7.16

In Fig. 7.16, both leaf and case are raised to the *same* potential, since they are connected by a wire and the case is insulated. Therefore *no divergence* of the leaf occurs.

Potential of a conductor

The potential at all points on the surface of a charged conductor is *the same*. If this were not so, electrons would flow from one point to another until it was.

Capacitance

The potential of a conductor changes when it is given a charge, and the larger the conductor, the smaller the change.

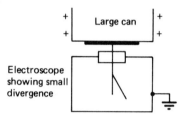

Figure 7.17

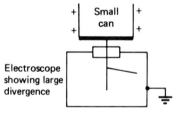

Figure 7.18

In Figs 7.17 and 7.18, with the same quantity of charge given to each can, the larger can is said to have a larger capacitance. It is able to store more charge before discharging. If the potential of a conductor changes by an amount V when it is given a charge Q, its capacitance C is given by

$$C = \frac{Q}{V}$$

The units of charge Q, potential V, and capacitance C are the **coulomb** (symbol C), the **volt** (symbol V) and the **farad** (symbol F) respectively. One farad is extremely large and so micro or pico farads are more commonly used in electronics.

Parallel plate capacitor

The potential of a charged conductor falls when another conductor, particularly an earthed one, is brought up to it. As the potential falls, the capacitance increases since the charge remains the same.

A capacitor stores charge and in a simple one two parallel metal plates are separated by an insulator. Its capacitance increases when:

(1) the separation of the plates decreases,
(2) the overlap of the plates increases,
(3) the insulator is a solid and not air. (The insulator is called the *dielectric.*)

Practical capacitors

Capacitors are used in radio, TV and computer circuits. They are not easy to make and their use is minimised. There are three main types.

(1) **Waxed paper**—consisting of two long metal strips of foil with waxed paper between them, rolled up (approx. 0.1 *μ*F).
(2) **Electrolytic**—a very thin coating of aluminium oxide is formed between two strips of aluminium foil (approx. 0.1 F). It is compact, but has to be charged in a certain sense and needs to be connected the right way around in a circuit.
(3) **Variable**—used for tuning radios and is made up of fixed and rotating plates separated by air (approx. maximum 500 pF). The plate overlap can be altered changing the capacitance.

Inverse square (or Coulomb's) law

The force acting between two *point charges* is inversely proportional to the *square* of their *distance* apart r and directly proportional to the product of their *charges* Q_1 and Q_2.

$$F \propto \frac{Q_1 Q_2}{r^2}$$

Particles such as electrons and protons are effectively point charges. (Compare this law to Newton's law of gravitation.)

In full, the Coulomb force F is given by

$$F = \frac{1}{4\pi\varepsilon}\left(\frac{Q_1 Q_2}{r^2}\right)$$

The constant of proportionality ε is called the *permittivity* of the medium which is between the point charges. The greater the permittivity of the medium the smaller the *Coulomb force* between the charges.

The *permittivity of free space* is the permittivity of a vacuum and has the symbol ε_0.

From Coulomb's law the units of permittivity are $C^2\,N^{-1}\,m^{-2}$ but this is simplified to *farads per metre.*

The value of ε_0 is taken as the permittivity of air since both are almost the same.

Electric field strength

An **electric field** is a region of force between charges and is represented by **lines of force**. The *direction* of the field at a point is the direction of the force on a *small positive charge*. The magnitude of the field is called the *electric field strength E.*

E at a point is the ratio of the force F on a small positive charge at that point to its charge Q_p

$$E = \frac{F}{Q_p}$$

Although the units of E from the equation are N C^{-1} the units usually used are *volts per metre*.

For a point charge Q_1, the force F acting between it and a small positive charge Q_p is

$$F = \frac{1}{4\pi\varepsilon}\left(\frac{Q_1 Q_p}{r^2}\right)$$

where r is the distance between charges.

Now $E = F/Q_p$ so the electric field strength for a point charge Q_1 is

$$E = \frac{Q_1}{4\pi\varepsilon r^2}$$

Electric potential

This is defined at a point as the work done per unit positive charge in moving it from *infinity* to that point. An infinite or very large distance from the charge is chosen as a *zero of potential*.

The potential due to a point charge Q at a distance r from it is

$$V = \frac{Q}{4\pi\varepsilon r}$$

Equipotentials

These are points in a field which have the same potential and form an *imaginary equipotential surface*. A charge moving over such a surface does *no work*. The equipotentials for a point charge lie on concentric spheres from its centre.

Field lines are always at right angles to equipotential surfaces as no force acts across the surfaces. A potential can be affected by the presence of other charges and these must be taken into consideration.

The work done when a charge moves between two points which have a difference in potential of V is given by

$$W = QV$$

For an electron passing through a p.d. of 1 V the energy change is

$$W = eV$$

where e is the *electronic charge*, and so

$$W = 1.6 \times 10^{-19} \times 1$$

$$W = \underline{1.6 \times 10^{-19}} \text{ J}$$

This quantity of energy is called the **electron-volt** (symbol eV).

Potential gradient

Figure 7.19

The force on a charge Q at A due to the electric field strength E is

$$F = QE$$

On moving the charge a short distance Δx from A to B the electric force does ΔW amount of work on the charge (Fig. 7.19):

$$\Delta W = F \Delta x$$

Therefore

$$\Delta W = QE \, \Delta x$$

For a p.d. of ΔV between A and B then

$$\Delta V = \frac{-\Delta W}{Q}$$

The negative sign is included since the higher the potential, the smaller the work needed to be done by the electric field.

$$\Delta V = \frac{-QE \, \Delta x}{Q}$$

As a result, the *field strength* E at the point A can be written as

$$\boxed{E = \frac{-\mathrm{d}V}{\mathrm{d}x}}$$

This is called the **potential gradient** in the x-direction. It means that the field strength E can also have units of *volts per metre*. For a *uniform field* (e.g. that between two parallel metal plates), E is *constant*. It is the same at all points (except at the very edges). For two plates at a p.d. V and separation d, the electric field strength E at all points is

$$\boxed{E = -V/d}$$

For the following assume that the permittivity of free space and the charge on an electron e are $8.85 \times 10^{-12} \, \mathrm{F\,m^{-1}}$ and $1.6 \times 10^{-19} \, \mathrm{C}$ respectively.

Example 7.1

Find (i) the electric potential and (ii) the electric field strength at a point X midway between two charges $+e$ and $-2e$ which are separated by 0.15 mm of air.

(i) Electric potential due to a point charge Q at a distance r from it is given by

$$V = \frac{Q}{4\pi\varepsilon_0 r}$$

Thus, the resultant potential at X is

$$V_x = \frac{e}{4\pi\varepsilon_0 r} + \frac{-2e}{4\pi\varepsilon_0 r} = \frac{-e}{4\pi\varepsilon_0 r}$$

$$V_x = \frac{-1.6 \times 10^{-19}}{4\pi \times 8.85 \times 10^{-12} \times 0.075 \times 10^{-3}}$$

Thus,

$$V_x = -2 \times 10^{-5} \text{ V}$$

(ii)   The field strength at a distance r from a charge Q is given by

$$E = \frac{Q}{4\pi\varepsilon_0 r^2}$$

Thus, the resultant electric field strength at X is

$$E_x = \frac{e}{4\pi\varepsilon_0 r^2} + \frac{-2e}{4\pi\varepsilon_0 r^2} = \frac{-e}{4\pi\varepsilon_0 r^2}$$

Therefore

$$E_x = \frac{-1.6 \times 10^{-19}}{4\pi \times 8.85 \times 10^{-12} \times 0.075^2 \times 10^{-6}}$$

Thus,

$$E_x = -0.256 \text{ V m}^{-1}$$

Example 7.2

Find the ratio of the electrostatic force between charges q and Q, to the gravitational force between them if they have charges 2.4×10^{-19} C and 3.8×10^{-19} C and masses 8.9×10^{-31} kg and 1.5×10^{-30} kg respectively. $G = 6.7 \times 10^{-11}$ N m^2 kg^{-2}.

The electrostatic force

$$F_e = \frac{qQ}{4\pi\varepsilon_0 r^2} \qquad \text{(from Coulomb's law)}$$

The gravitational force

$$F_g = \frac{GmM}{r^2} \qquad \text{(from Newton's law)}$$

Thus,

$$\frac{F_e}{F_g} = \frac{qQ}{4\pi\varepsilon_0 mMG} = \frac{2.4 \times 10^{-19} \times 3.8 \times 10^{-19}}{4\pi \times 8.85 \times 10^{-12} \times 9 \times 10^{-71}}$$

Therefore

$$F_e/F_g = 9.1 \times 10^{42}$$

Thus, for small masses, the gravitational force is considerably weaker than the electrostatic force. It only becomes significant when dealing with astronomical bodies.

Equivalence of static and current electricity

Two experiments can be used to demonstrate the equivalence of static and current electricity.

Experiment 1

See Fig. 7.20.

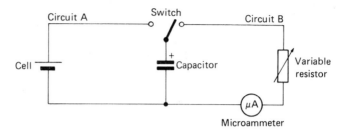

Figure 7.20

When the capacitor is switched to circuit A, electrons flow from the cell to the negative plate of the capacitor. Eventually a positive charge builds up on the other plate by *induction*, and a potential difference exists between the plates. When the p.d. across the plates is equal to the p.d. across the cell but is in the *opposite sense*, no more current flows. The capacitor is *fully charged* with charge $Q = CV$.

On switching to circuit B, there is no opposing p.d. to that of the capacitor and it begins to *discharge* (i.e. it begins to act like a cell, but the p.d. across it falls to zero when all the positive charge on one plate has been neutralised by electrons which have been attracted to it from the other).

As the capacitor is discharging, the *microammeter* registers a current I which is kept *constant for as long as possible* (time t) by adjusting the variable resistor. Now $Q = It$ should give the same charge as was obtained from the first part of the experiment.

Experiment 2

See Fig. 7.21.

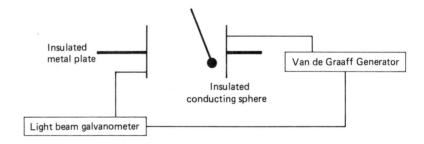

Figure 7.21

In this experiment, when the Van de Graaff generator is switched on the insulated conducting sphere moves from plate to plate. On touching a plate it becomes charged with the same charge as is on the plate and is repelled, to the other plate. A similar thing occurs at the other plate.

A transfer of charge therefore occurs every time the sphere touches a plate, and this is registered on the galvanometer, showing that a small current has been produced.

Current electricity

When a material is a **good conductor**, it has many *free* or *conduction electrons*. In a conductor the electrons which are 'free' move randomly with a speed of about 3×10^5 m s^{-1}. As the motions are *random* there is no *total* current.

An *electric field* in a conductor (set up by a battery for example) causes the free electrons to *accelerate*. As they accelerate they gain kinetic energy and pass this onto any atoms they collide with. On a macroscopic scale this gives rise to *temperature*.

The *total acceleration* of the free electrons is *zero* because of these collisions and they acquire a constant *drift velocity*.

Charge is a property of a molecule or particle which creates a *force* between it and another which also has charge, e.g. electrons and ions. When electric charges flow through a material a current is created. The material is a **conductor**. If no current passes it is an **insulator**.

Electric current

An ion is an electrically *charged atom* or *molecule*. The atom or molecule has acquired an *excess* or *deficiency* of electrons. When a chemical compound forms ions (**ionises**) in solution or when molten, it is called an **electrolyte** if wires and a battery connected to it give a continuous current and a *chemical change* occurs.

For example mercury is *not* an electrolyte since, although it conducts current electricity, there is no resulting *chemical change*.

A closed path which allows a current is a **circuit**. The battery has a surplus of electrons at one terminal and a deficit of them at the other. A current tries to even this out, but the battery maintains its initial condition until it dies.

Effects of a current

(1) **Heating effect**—due to energy lost in *collisions* of charged particles with the atoms of the conductor.

(2) **Chemical effect**—due to movement of ions in different directions during *electrolysis*.

(3) **Magnetic effect**—Any moving charged particle has a **magnetic field**, and this can be seen by the deflection of a compass needle near a wire carrying current. (An electron moving around an atom creates a magnetic field.) The charged particle also carries an *electric field*.

Current

Current I is the quantity of charge Q passing a given point in time t.

$$I = \frac{Q}{t}$$

The *direction* of current is by *convention* that which a positive charge would take at the point in the circuit.

Ampère

An ampère (A) is that current which, when passing through two infinitely long, parallel straight wires of negligible cross-section, 1 m apart in a vacuum, produces a force between them of $2 \times 10^{-7} \, \text{N m}^{-1}$.

Coulomb

A coulomb (C) is the charge passing a given point in a circuit when a current of 1 A flows for one second.

The charge on an electron is very small and has the symbol e.

$$e = 1.6 \times 10^{-19} \, \text{C}$$

Therefore

$$\boxed{1 \, \text{C} = 6.25 \times 10^{18} e}$$

Drift velocity of electrons

A conductor of length L and cross-section A has n free electrons per unit volume each with charge e (see Fig. 7.22).

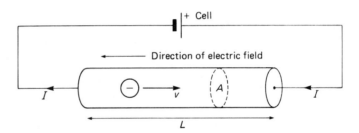

Figure 7.22

volume of conductor $= AL$

number of electrons in that volume $= nAL$

So

total charge of that volume, $Q = nALe$

When a battery is put across the conductor,

$$I = \frac{Q}{t} = \frac{nALe}{t}$$

drift velocity of an electron, $v = L/t$

Therefore

$$\boxed{I = nAve}$$

Magnitude of electron drift velocity

Electron drift velocities usually turn out to be *very small*. The current $I = nAve$ can be large though, because n, the *electron density*, is very high. For example, for copper $n = 1 \times 10^{29} \, \text{m}^{-3}$. So for a wire carrying 1 A and cross-section 2×10^{-7} m^2,

$$v = \frac{I}{nAe} = \underline{3.1 \times 10^{-4} \, m \, s^{-1}}$$

In an *electrolyte* the *cross-section* is usually greater, and the drift velocity smaller. In a vacuum there is nothing to impede electron movement and one-quarter of the speed of light can be achieved by some electrons.

Electrical energy

Current in a circuit can give rise to other forms of energy, e.g. heat and mechanical energy. Other forms of energy can be converted back into electrical energy.

dynamo—mechanical energy to electrical energy

battery—chemical energy to electrical energy

Electrical energy is due to *forces* acting on charged particles producing a current. A battery or dynamo is said to produce an **electromotive force** (emf).

The emf of a source (battery, dynamo, etc.) is the *energy converted* into electrical energy when unit charge passes through it.

Potential difference (p.d.)

When a source of emf supplies energy to a circuit there is an electric field between its terminals. The electric field acts inside the conductors in the circuit, setting charges in motion. The field does work on these electrons causing electrical energy to be converted into *heat*.

The field is described by giving *points* in the circuit a figure called the **potential**. Electrons flow from a *low* to a *higher* potential. The difference of potential between two points indicates the strength of the field driving the electrons from one to the other. (Earth is regarded as *zero potential*.)

The p.d. is the work done by the electric field per unit charge passing between two points.

$$\text{potential difference, } V = \frac{W}{Q} = \frac{W}{It}$$

(therefore 1 volt = $1 \, J \, C^{-1}$).

p.d. and emf

The **emf** drives the current around the whole circuit, including the source. The **p.d.** across any part of the circuit drives the current around that part.

When a current passes, some energy produced by a source of emf is lost in heat by driving the current through the source itself. (The source has *internal resistance*.) Therefore when a current passes (**closed circuit**), the p.d. across the terminals is less than the emf. When no current passes (**open circuit**), the p.d. across the terminals is equal to the emf.

The unit of both emf and p.d. is the **volt** (symbol V).

One volt is the p.d. between two points when the energy converted from electrical to other forms is one joule per coulomb that passes between them.

The **electric field** has the symbol E and is a vector quantity of magnitude given by V/L ($V m^{-1}$), where L is the length of conductor across which there is a p.d. V.

Water circuit

A comparison between a domestic water circuit and an electrical one is shown in Fig. 7.23.

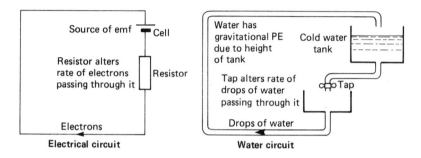

Figure 7.23   Electrical circuit   Water circuit

Resistance and Ohm's law

The p.d. across a metallic conductor which is not the site of an emf is directly proportional to a steady current flowing through the conductor, *provided* the temperature and other physical conditions remain constant.

$$V \propto I$$
$$V = RI$$

The constant of proportionality is the **resistance** R of the material of the conductor and has the unit the **ohm** (symbol Ω). The reciprocal of resistance is called the **conductance** G and has the unit the **siemens** (symbol S).

Exceptions to Ohm's law

(1) Any device which is the site of an emf does not obey Ohm's law, e.g. electric cell, diode valve.
(2) Some crystals, e.g. silicon carbide.
(3) Some non-metallic conductors, e.g. selenium carbide.
(4) Gases, e.g. as in the neon lamp.
(5) Some conducting solutions, e.g. copper-II sulphate solution.
(6) Conductors with high temperature changes, e.g. bulb filament.

Substances or devices which do not obey Ohm's law are called **non-ohmic.**

Internal resistance of a cell

The difference between the emf and the terminal p.d. of a cell when current is taken from it (sometimes called the 'lost volts') is due to the *internal resistance* of the cell r (Fig. 7.24).

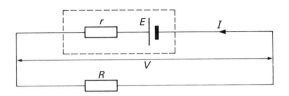

Figure 7.24

It is found that $E - V \propto I$ and so

$$\boxed{E - V = Ir}$$

Since $V = IR$,

$$\boxed{E = I(R + r)}$$

For a circuit $E = V$ only if $I = 0$ (i.e. on *open circuit*).

A **voltmeter** measures p.d. If it has a *high resistance* the current through it is small, and when it is placed across a cell it will measure the emf.

Power dissipation in resistors

Power is defined as the **rate of doing work**, and p.d. is defined as the **work done per coulomb in moving it between two points.** Therefore the work done is $W = VQ$ and the power is

$$\frac{W}{t} = \frac{VQ}{t} = VI$$

so

$$\boxed{P = VI} \quad \text{(unit: the watt)}$$

Combinations of resistors

Resistors in series

Figure 7.25

In series

The *total p.d.* across the resistors (Fig. 7.25) is

$$V_1 + V_2 + V_3$$

The *total current* through the resistors is

$$I_1 = I_2 = I_3 = I \quad \text{(say)}$$

So

$$IR_{\text{total}} = IR_1 + IR_2 + IR_3$$

and therefore

$$\boxed{R_{\text{total}} = R_1 + R_2 + R_3}$$

Resistors in parallel

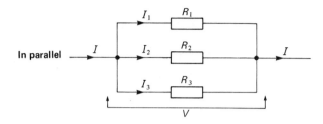

Figure 7.26

Now
$$I = I_1 + I_2 + I_3$$
The p.d. *V* is the *same* across each resistor (Fig. 7.26). So,

$$\frac{V}{R_{\text{total}}} = \frac{V}{R_1} + \frac{V}{R_2} + \frac{V}{R_3}$$

therefore

$$\boxed{\frac{1}{R_{\text{total}}} = \frac{1}{R_1} + \frac{1}{R_2} + \frac{1}{R_3}}$$

Resistivity

The resistance of a conductor depends upon its size and upon the material it is made of. For a *uniform conductor* it is directly proportional to the length and inversely proportional to the cross-sectional area.

$$\boxed{R = \rho\frac{L}{A}}$$

The constant of proportionality ρ is the **resistivity**. The resistivity of a certain material is the resistance of unit length and unit cross-sectional area of the material (units: $\Omega\,\text{m}$).

Conductivity σ is the reciprocal of resistivity and has units $\text{S}\,\text{m}^{-1}$.

Addition of impurities

Metals—resistivity is considerably increased by the addition of even small amounts of impurity, since it affects the movement of conduction electrons.

Semiconductors—e.g. germanium and silicon. These contain only a few conduction electrons, but with impurities the number is increased and so the resistivity is reduced.

Increase in temperature

Metals—when the temperature of a metal is increased, atomic vibrations increase, impeding even more the flow of electrons, and so resistivity is increased.

Semiconductors—the increased atomic vibrations free even more electrons, and so resistivity is reduced again. (In an *electrolyte* the ions are able to move more freely and resistivity is also reduced.)

Example 7.3

A series circuit consists of a 6 Ω resistor, ammeter and battery. The current in the circuit is 0.5 A but drops to 0.3 A when another resistor of 8 Ω is added in series. Find the battery emf and its internal resistance.

Applying the circuit equation

$$E = I(R + r)$$

then initially

$$E = 0.5(6 + r) = 3 + 0.5\,r$$

After addition

$$E = 0.3(14 + r) = 4.2 + 0.3r$$

Therefore

$$3 + 0.5r = 4.2 + 0.3r$$

$$0.2r = 1.2$$

Thus,

$$r = \underline{6\,\Omega}$$

Finally, the emf

$$E = 0.5(6 + 6) = \underline{6\,V}$$

Example 7.4

An electric heater consists of two wire coils each of length 15 m and 0.3 mm radius. Determine the power output for each of the possible combinations of the coils, if the device is used on a 160 V supply. (The resistivity of the coil wire is $8.2 \times 10^{-7}\ \Omega\,m$.)

Applying the equation

$$R = \frac{\rho L}{A} = \frac{\rho L}{\pi r^2}$$

then

$$R = \frac{8.2 \times 10^{-7} \times 15}{\pi \times (0.3 \times 10^{-3})^2}$$

Thus,

$$R = 43.5\,\Omega$$

For one coil on its own,

$$P = V^2/R = 160^2/43.5 = \underline{589\,W}$$

For two coils in series,

$$P = V^2/(R + R) = 160^2/(43.5 + 43.5) = \underline{294\,W}$$

For two coils in parallel,

$$\text{total resistance, } R_{\mathrm{p}} = \frac{R \times R}{R + R} = \frac{43.5 \times 43.5}{43.5 + 43.5} = 21.8\,\Omega$$

Therefore

$$P = V^2/R_{\mathrm{p}} = 160^2/21.8 = \underline{1174\,W}$$

Example 7.5

A circuit is set up as shown in Fig. 7.27 using a cell of emf 1.6 V and negligible internal resistance. What percentage of the p.d. indicated by the voltmeter when the switch is open will be given when the switch is closed, assuming that the voltmeter has a resistance of 40 Ω?

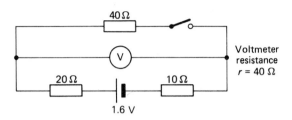

Figure 7.27

When the switch is open the total resistance is

$$R_s = 40 + 20 + 10$$

When the switch is closed the total resistance is

$$R = 20 + 20 + 10$$

Fraction of p.d. is

$$\frac{I'r/2}{Ir} = \frac{Vr/2R}{Vr/R_s} = \frac{(1.6 \times 40)/(2 \times 50)}{(1.6 \times 40)/70} = 0.7$$

i.e.   percentage p.d. = 70%

Temperature coefficient of resistance, α

This is the *fractional increase* in a material's resistance at 273 K per unit rise in temperature.

$$\boxed{\alpha = \frac{R_T - R_{273}}{R_{273}(T - 273)}} \quad \text{(unit: } K^{-1})$$

R_T is the resistance of the material at temperature T (in *kelvin*). (A negative coefficient indicates resistance decreases with rise in temperature.)

Example 7.6

A nickel coil is heated from a temperature of 293 K to 333 K. What is its final resistance if it initially was 10 Ω? (For nickel $\alpha = 4.3 \times 10^{-3}$ K^{-1}.)

$$R_{293} = R_{273}(293 - 273) + R_{273}$$
$$R_{333} = R_{273}(333 - 273) + R_{273}$$

Dividing the bottom equation by the top one,

$$\frac{R_{333}}{R_{293}} = \frac{60\alpha + 1}{20\alpha + 1}$$

By substituting the value for the resistance at 293 K,

$$R_{333} = \underline{11.6\,\Omega}$$

172   *Concise Physics*

Types of resistor

Fixed resistor

Symbol ⊣▭⊢ Usually contains a mixture of carbon and non-conducting material.

Variable resistor

Symbol ⊣▱⊢ Usually circular with rotating arms that slide around in a broken circle, contact.

Solenoid wire wound variable resistor

This consists of a cylindrical wire coil, along which a contact can slide. Variable resistors are used in two ways:

(1)   as a **rheostat** controlling current in a low resistance device (Fig. 7.28),

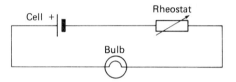

Figure 7.28

(2)   as a **potential divider** controlling the p.d. across a device (Fig. 7.29).

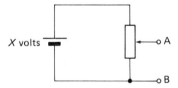

Figure 7.29

(Note: The p.d. across A and B can be varied from 0 V to XV.)

Potentiometer

This is used to measure p.d. *accurately* or, if adapted, current and resistance.

A battery is joined to the ends of a resistance wire so that its p.d. is applied across the whole length of the wire. *Any fraction* of this p.d. can be tapped off between a probing contact ('jockey') and one end of the wire. This instrument is used for *d.c. only* and the battery is often referred to as the **driver cell**. A *centre-reading galvanometer* connected to the probing contact indicates the current passing through the length of resistance wire tapped off. Galvanometers are devices for *detecting* or measuring small currents. They are not usually given scales in ampères and need to be *calibrated* for measurements.

The potentiometer and its uses

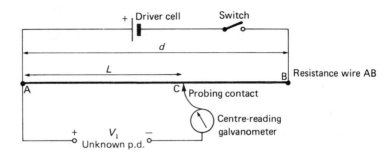

Figure 7.30

See Fig. 7.30. The *p.d. tapped off* between A and C acts in an anticlockwise direction but V_1 acts in the opposite direction. If these two p.d.'s are *equal*, no current flows through the galvanometer, and the potentiometer is said to be *balanced*. (Off balance the galvanometer needs a protective resistance.)

At balance the distance L is measured and the unknown p.d. = p.d. tapped off, so

$$\boxed{V_1 = I \times Lr}$$

where r is the resistance of wire per unit length and I is the current from the driver cell.

If the p.d. across the whole wire is V and the distance AB is d, then

$$V = I \times dr$$

Then $\dfrac{V_1}{V} = \dfrac{ILr}{Idr}$

Therefore

$$\boxed{V_1 = \frac{L}{d} V}$$

Advantages

(1)   The potentiometer scale depends on the length d and therefore by increasing d the *sensitivity* can be increased. For practical purposes the length is made to be one metre.

(2)   At balance the galvanometer gives a *null reading* indicating no current and *no measurement* of current is required.

(3)   At balance no current is drawn from the circuit being tested and so V_1 is *unaltered*.

(4)   The measurements can be made with reference to a *standard cell* (i.e. a cell whose emf is accurately known and which maintains this emf constant over a long period).

Disadvantages

(1)   Cumbersome.

(2)   Slow to use.

Practical points

(1)   Galvanometer deflections should be in *opposite directions* when the probing contact is placed at the ends of the wire. If this is not so, the unknown p.d. is connected the *wrong way* around or the driver cell emf is less than the unknown p.d.

(2)   The probing contact must *not be drawn* across the wire or the wire will become thinned in places and so a probing action is used.

(3)   For accuracy the length L should be made as long as possible, i.e. about half the length of the potentiometer wire.

Comparison of emf's of two cells

If a cell replaces the unknown p.d. then emf of cell = p.d. across AC at balance. If the measurement is repeated using another cell then a comparison can be made.

Comparison of emf's

When the balance lengths for two cells of emf E_1 and E_2 are obtained on a potentiometer as L_1 and L_2 respectively, then

$$E_1 = \frac{L_1 V}{d} \quad \text{and} \quad E_2 = \frac{L_2 V}{d}$$

where V/d is the p.d. per unit length across the potentiometer wire as applied by the driver cell. So

$$\boxed{\frac{E_1}{E_2} = \frac{L_1}{L_2}}$$

A **Weston standard cell** keeps a constant emf of 1.0186 V (at 293 K) when the current taken from it is less than ten micro-ampères. To do this a resistor of value 0.5 MΩ is put in series with it. If the Weston cell is one of those being compared, the emf of the other cell can be found accurately and the potentiometer is effectively *calibrated*. (The high resistance has to be shorted out near balance point.)

The calibration is the p.d. across unit length of the wire and is given by $1.0186/L^*$, where L^* is the balance length for the Weston standard cell. Other standard cells can be used in the same way.

Note: A standard cell can be used to see whether an accumulator has run down at the end of the experiment by re-checking the balance point.

Calibrating a voltmeter

The potentiometer is *calibrated first*, using a standard cell to establish the p.d. per unit length of the wire provided by the driver. An accumulator and a variable resistor are connected to form a *potential divider*, enabling various potential differences to be applied across the voltmeter to be calibrated and across the potentiometer wire. The potential divider is connected in parallel with wire between points P and Q as shown in Fig. 7.31. Readings of balance lengths are taken by varying the potential

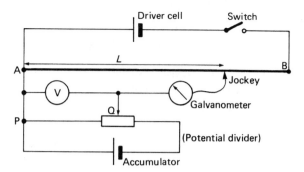

Figure 7.31

difference across PQ and the p.d. values calculated from the potentiometer calibration.

These values can then be *marked* on the voltmeter scale.

Measurement of thermoelectric emf

A suitably adapted and calibrated potentiometer is able to measure the *very small* emf's produced by a thermocouple.

Calibration circuit

See Fig. 7.32.

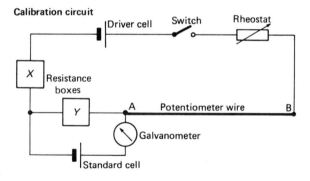

Figure 7.32

Thermocouple circuit

See Fig. 7.33.

Experiment to determine thermoelectric emf

Thermoelectric emf's are of the order of mV and therefore with an ordinary potentiometer the balance lengths would be too short to measure. To make the instrument more *sensitive* its length has to be increased. In practice this is achieved by putting a large resistance (*call it Y*) *in series* with the potentiometer wire.

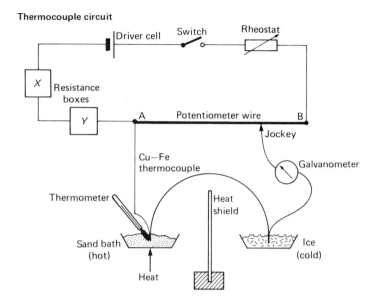

Figure 7.33

Calibration

In order to determine the thermoelectric emf for a given balance length, the potentiometer needs to be calibrated.

Using a suitable meter the *resistance* of the potentiometer wire (*call this r*) is found. The value of the resistance Y is now set to $102r$ and the complete calibration circuit set up as shown in Fig. 7.32.

By varying the resistance of the resistance box X and making fine adjustments using the rheostat, a zero deflection is obtained on the galvanometer, remembering that the galvanometer is connected to end A of the potentiometer wire.

When this balance is achieved, the p.d. across the resistance Y must be *equal* to the emf of the standard cell (Weston-cadmium) placed in parallel ·with it as shown.

Thus, the p.d. across each r unit of resistance must be given by

$$V_r = \frac{1.02}{102} = \underline{0.01 \text{ V}}$$

Since both Y and the potentiometer wire are in series, the p.d. across the wire (which has a measured resistance of r) must also be 0.01 V. Thus, the potentiometer is calibrated and can measure thermoelectric emf's up to a value of 0.01 V or 10 mV.

Measurements

To make measurements, all the resistance settings obtained in the calibration are left *unaltered* and the thermocouple circuit set up as shown in Fig. 7.33. The sand bath is gradually heated, with balance lengths being carefully recorded at regular intervals of temperature. Readings should be repeated as the sand bath cools. Thermoelectric emf's corresponding to different temperature differences across the thermocouple junctions can then be calculated using the potentiometer calibration. A plot of emf versus temperature should reveal the characteristic parabolic curve, provided the neutral temperature has a fairly low value.

Simple cell

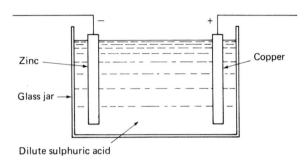

Figure 7.34

The simple cell (Fig. 7.34) has an emf of about 1 V but stops working after a time due to *polarisation*. This is the collection of gas bubbles on the copper plate. Potassium dichromate can depolarise the cell by oxidising the hydrogen gas to give water. Impurities in the zinc cause *local action* which results in the zinc being used up—even when current is not being supplied by the cell. The simple cell cannot be recharged and is called a *primary cell*. It is no longer widely used.

Using a potentiometer to calibrate an ammeter

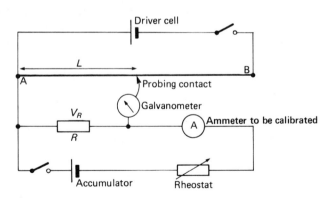

Figure 7.35

The potentiometer is *calibrated* with a *standard cell* to establish the p.d. across unit length. The p.d.'s across a standard resistor R are then obtained from balance lengths as the current through R supplied by an accumulator is varied by means of a *rheostat* connected in series with it, as shown in Fig. 7.35. The ammeter to be calibrated is also put in series with R, the value of which is chosen for a *large balance length*.

Comparison of resistances

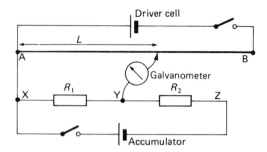

Figure 7.36

The current supplied by the accumulator is the same in R_1 and R_2 (Fig. 7.36). So $V_1 = IR_1$ and $V_2 = IR_2$. The balance lengths across *each* resistor are found in turn, i.e. for the p.d. across XY *and then* for the p.d. across YZ. If V/d is the p.d. per unit length of the wire, then

$$V_1 = IR_1 = L_1 \frac{V}{d} \quad \text{and} \quad V_2 = IR_2 = L_2 \frac{V}{d}$$

Therefore

$$\boxed{\frac{R_1}{R_2} = \frac{L_1}{L_2}}$$

If the two respective balance lengths L_1 and L_2 are to be large, V_1 and V_2 should be near V_{AB}. Should R_1 and R_2 overheat, a smaller current is used and a suitable resistance put in series with the wire AB as was done in the measurement of thermoelectric emf by the potentiometer, to make L_1 and L_2 large again.

Determining the internal resistance of a cell

The potentiometer can be used to find the *internal resistance* of a cell. In this case a circuit containing the cell and a known resistance R with a switch are put *in series* with the galvanometer (Fig. 7.37).

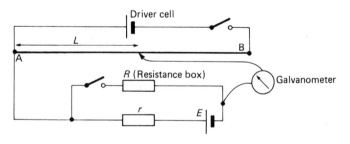

Figure 7.37

The cell in Fig. 7.37 has been represented by a d.c. source and a resistor r. With the switch in series with R open, the p.d. across the cell at balance length L is its emf E. A known resistance R from a resistance box is then switched in and a new balance length L_R found for the p.d. applied across R by the cell.

Further values of R are taken and in each case the new balance length L_R found.

If V/d is the p.d. across unit length of the wire, then for the first balance length L the emf of the cell is

$$E = L\frac{V}{d}$$

(but $E = I(R+r)$ also).

For the second balance length L_R (and all the other values taken) the p.d. across the resistor R is

$$V_R = L_R\frac{V}{d}$$

(but $V_R = IR$ also).

Therefore

$$\frac{E}{V_R} = \frac{I(R+r)\cdot}{IR} = \frac{L}{L_R}$$

Rearranging the last part of the above equation gives

$$r = R\frac{L}{L_R} - R$$

Further rearrangement gives

$$\frac{1}{R} = \frac{L}{rL_R} - \frac{1}{r}$$

The purpose of this rearranging has been to obtain an equation of the form $y = mx + c$ which incorporates in some way the different values of R and L_R that the experiment has given.

A graph of $1/R$ versus $1/L_R$ will give a *straight line* whose gradient is given by L/r and which has a negative intercept on the y-axis of $1/r$ (Fig. 7.38).

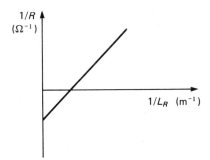

Figure 7.38

Example 7.7

A potentiometer circuit is set up as shown in Fig. 7.39. With the switch open and then closed, the balance lengths are 82.3 cm and 75.8 cm respectively. If r is the internal resistance of the cell, find the value of r.

The emf of the cell $E \propto L_1$, and the p.d. across the cell and the resistor $V \propto L_2$, where L_1 and L_2 are the balance lengths respectively. Thus,

$$E/V = L_1/L_2$$

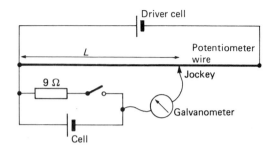

Figure 7.39

But also

$$\frac{E}{V} = \frac{I(R+r)}{IR} = 1 + \frac{r}{R} = \frac{L_1}{L_2}$$

Therefore

$$r = R(L_1/L_2) - R = (9 \times 82.3/75.8) - 9$$

Thus,

$$r = \underline{0.8\,\Omega}$$

Example 7.8

A resistance box X and 1.5 V accumulator are placed across a potentiometer wire as shown in Fig. 7.40. If the potentiometer wire is 1.5 m long and has a resistance of 7 Ω, find the value of X which will produce a p.d. of 2 μV per mm along the wire.

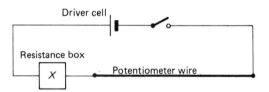

Figure 7.40

Assuming that the connecting wires have negligible resistance, then the total resistance of the circuit is $X + 7$ and since the circuit is a series one, the p.d. across the wire must be in proportion to the p.d. across resistance box X.

$$\text{p.d. across whole wire, } V_w = 2 \times 10^{-6} \times 1.5 \times 10^3 = 3 \times 10^{-3}\,\text{V}$$

But

$$V_w = \frac{7V'}{X+7}$$

where $V' =$ accumulator p.d. Thus,

$$X = \frac{7V'}{V_w} - 7 = \underline{3493\,\Omega}$$

Back emf

When an electrode becomes covered in a layer of gas, a so-called *back emf* acts in opposition to the p.d. applied across the voltmeter and reduces it as a result.

Due to this *polarisation effect* the current through the voltmeter is given by

$$I = \frac{V - E_b}{R}$$

where E_b is the *back emf* due to the voltmeter, V is the p.d. applied across the voltmeter, and R is its resistance.

A graph of I versus V is shown in Fig. 7.41.

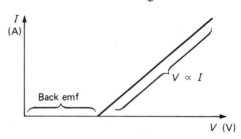

Figure 7.41

The graph shows that the voltmeter is acting as a *non-ohmic device*.

A back emf is obtained when copper-II sulphate solution is the electrolyte and *inert electrodes* are used (carbon or platinum). (No back emf is exhibited when the electrodes are copper.)

Faraday's constant, *F*

This is the charge carried by a mole of any monovalent ions (i.e. each ion has a charge of $+1$ or -1), and is approximately $96\,500\,\text{C}$.

number of ions in a mole = Avogadro's constant $N = 6 \times 10^{23}$

Therefore each ion carries a charge of

$$\frac{96\,500}{6 \times 10^{23}} = 1.6 \times 10^{-19}\,\text{C}$$

i.e. $Ne = F$

where e = the charge on one electron.

Charge to mass ratio of ions

From **Faraday's first law** of electrolysis, the mass of substance deposited during electrolysis m is

$$m = zQ$$

where Q is the *charge passed* and z is the *electrochemical equivalent*. If each ion carries a charge ne, where n is the *valency* of the ion, and has mass m_i, then for N ions discharged at an electrode

$$Nm_i = zNne$$

and

$$\frac{ne}{m_i} = \frac{1}{z}$$

8
Magnetism and a.c. theory

Introduction

This chapter is basically a continuation of the work on electricity. It begins with a look at natural magnetism and how magnetic materials can be magnetised. The domain theory and the use of field lines to represent magnetic fields follow. Different types of field, particularly those due to current carrying conductors, are then shown. This leads to a discussion of the force on such conductors in magnetic fields and the idea of magnetic flux density. Such ideas are used to define the ampère and form the basis of the simple current balance. This is supported by an investigation of the effect of a magnetic field on a rectangular current carrying coil and its application to meters. The next section concerns electromagnetic induction and the laws governing it. Its application to the generation of electricity is also noted.

A gradual introduction to alternating current is given in a discussion of the properties of transformers, which precedes detailed sections on two major a.c. components—the inductor and the capacitor. The last part of the chapter considers a.c. properties and the effect of a.c. upon different types of component. Particular note is made of the concept of phase difference between supply p.d. and the p.d.'s across components. It finishes with a look at resonant circuits and a survey of the cathode ray oscilloscope, which is able to measure both steady and alternating p.d.'s.

Magnetism revision targets

The student should be able to:

(1) Discuss the basic properties and facts of natural magnetism.
(2) Graphically represent examples of magnetic field.
(3) State and apply Fleming's left hand rule to current carrying conductors and moving charges in uniform magnetic fields.
(4) Describe the factors affecting the force on a current carrying conductor in a magnetic field.
(5) Define magnetic flux density and apply it to numerical problems.
(6) Discuss how magnetic flux density varies around a straight wire conductor and a solenoid.
(7) Apply the equations $F = BIL$ and $F = BQv$ to numerical problems.
(8) Define the ampère and explain the basis of the simple current balance.
(9) Determine the moment of the couple causing a rectangular coil to rotate in a uniform magnetic field, when it carries current.
(10) Describe the basic principle of the moving coil galvanometer.

(11)   Calculate the values of resistors which need to be connected to the moving coil galvanometer for it to read different scale ranges.
(12)   Discuss how the sensitivity of the meter may be affected.
(13)   Explain and distinguish between induction and mutual induction.
(14)   Define the terms magnetic flux and flux linkage.
(15)   State and give the equation for Faraday's law of electromagnetic induction.
(16)   State Lenz's law and describe how it can be experimentally investigated.
(17)   Give Fleming's right hand rule and distinguish it from his left hand rule.
(18)   Explain what is meant by an eddy current.
(19)   Discuss the induced emf across a coil rotating in a uniform magnetic field and its application to generators.
(20)   Define the term back emf.
(21)   Explain the working of a transformer and distinguish between a step-up and a step-down transformer.
(22)   Give the transformer equation and apply it to numerical problems and a consideration of energy losses.
(23)   Define and give the equations for inductance and mutual inductance.
(24)   Define the term capacitance and discuss the factors which affect the capacitance of a parallel plate capacitor.
(25)   Solve numerical problems involving parallel plate capacitors.
(26)   Explain what is meant by permittivity and discuss how relative permittivity can be experimentally determined.
(27)   Describe the operation of an electrometer/d.c. amplifier and how it can be used to determine the charge stored on a capacitor.
(28)   Perform calculations involving simple parallel and series networks of capacitors.
(29)   Calculate the energy stored in a capacitor and describe the charging and discharging characteristics of a capacitor.

Alternating current revision targets

The student should be able to:

(1)   Explain what is meant by alternating current.
(2)   Explain the purpose of a diode and how it can be used to rectify alternating current.
(3)   Define the term rms value and obtain the relationship between rms value and peak value.
(4)   Discuss the effect of a.c. on a capacitor.
(5)   Explain the significance of phase with respect to components on a.c.
(6)   Define and distinguish between reactance and impedance.
(7)   Use the expressions for reactance and impedance in the quantitative analysis of a.c. circuits.
(8)   Explain what is meant by, and calculate for a given circuit, the resonant frequency.
(9)   Describe the application of resonance to tuning circuitry.
(10)   Give a functional description of the various parts of a cathode ray oscilloscope.
(11)   Describe the nature of the time base.

(12)   Explain how the oscilloscope can be used to determine both steady and alternating p.d.'s.

(13)   Discuss the advantages of the oscilloscope over conventional instruments.

(14)   Incorporate the oscilloscope in experiments and circuits discussed throughout electricity and magnetism topics.

Properties of magnets

Magnetic materials

Only certain substances are attracted strongly to magnets, e.g. iron, steel and nickel. These are **ferromagnetic**.

Magnetic poles

These are regions to which magnetic materials are attracted, and always occur in pairs of equal strength.

North and south poles

A suspended magnet points with one pole (N-seeking) towards the Earth's north, and therefore can be used as a compass.

Law of magnetic poles

Like poles repel, unlike poles attract.

Test for a magnet

An unmagnetised magnetic material attracts both poles of a magnet which has been suspended so that it can swing freely. A permanent magnet repels a suspended magnet with one of its poles.

Methods of magnetisation

Stroking

There are two methods of magnetising by stroking. It is possible to use either one permanent magnet (single touch) or two (double touch) in the process. See Fig. 8.1.

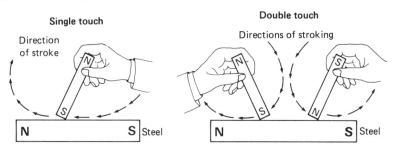

Figure 8.1

For single touch, steel is stroked many times always in the same direction with the same pole of one magnet. Weak magnets are produced. For double touch, the unlike poles of two magnets are applied at the same time in opposing directions. The magnets must be lifted high above the steel at the end of each stroke in both methods. The pole at the end of a stroke is opposite to the stroking pole.

Electrically

A cylindrical coil called a solenoid is connected to a *d.c. supply,* and the magnetic material put inside it. A current passed through the coil renders the material magnetised and the direction of the poles depends on the direction of the current.

Right hand grip rule
If the fingers of the right hand are wrapped around the coil in the direction in which the current is passing, then the thumb of that hand points towards the N-seeking pole of the magnet formed in the coil. See Fig. 8.2.

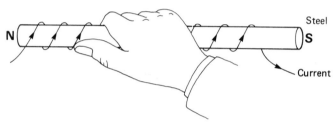

Figure 8.2

Induced magnetism

Magnetism is induced in unmagnetised magnetic materials by bringing the material near to the pole of a permanent magnet. See Fig. 8.3.

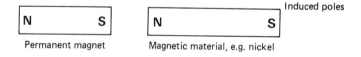

Figure 8.3

Magnetic properties of iron and steel

Soft iron becomes a strong magnet in a magnetic field, but loses the magnetism when there is *no field*. It is used as the core of an **electromagnet**. Steel is more difficult to magnetise, but remains a magnet when the field is removed and is used to make permanent magnets.

Theory of magnetism (domain theory)

A magnet can be thought of as an assembly of very small magnets all lined up with their N-seeking poles pointing the same way. At the ends the 'free' poles of the tiny magnets, called domains, repel each other. See Fig. 8.4.

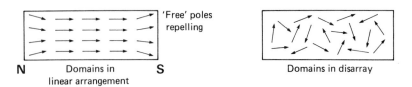

Figure 8.4

In an unmagnetised bar the domains are in a state of disarray. The N-pole of one is neutralised by the S-pole of another, and there are no free poles at the ends.

Magnetic saturation

This is the limit to the strength of a magnet. It arises when all the domains have been aligned.

Methods of demagnetisation

(1)   By **heating** or **hammering:** violent disturbance of the domains causes them to be in disarray again.

(2)   By **electrical means:** an *a.c. supply* is used. The magnet is slowly withdrawn from a solenoid carrying the a.c. to a good distance away, and the supply switched off. See Fig. 8.5.

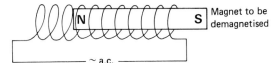

Figure 8.5

Storing magnets

Magnets become weaker with time. The free poles near the ends repel each other and destroy the domain alignment. To stop this, **keepers** are used. These become induced magnets and their poles neutralise the poles of the bar magnets, as the domains form complete chains. See Fig. 8.6.

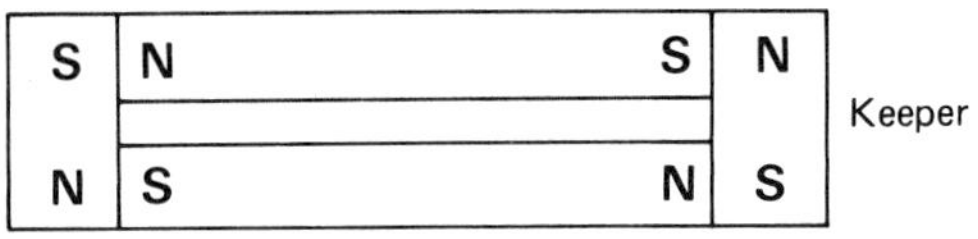

Figure 8.6

S	N		S	N
N	S		N	S

Keeper

Magnetic field

This is a space where a force acts (a) on a small suspended magnet or (b) on a moving charge.

Magnetic fields and lines of force

A magnetic line of force shows the path an *imaginary* free N-pole would take if placed at that point.

(1) A line of force goes from a N-pole to a S-pole.
(2) Lines of force *repel* each other sideways.
(3) They always try to shorten.
(4) If they are close together the field *is strong* in that region.
(5) An arrow indicates *the direction of the field*, i.e. the path an imaginary free N-pole would take.

Lines of force can be plotted for different magnets using plotting compasses—which are useful for weak fields—or by using iron filings. See Figs 8.7, 8.8 and 8.9.

Neutral point

A neutral point (marked X on diagrams) is where the resultant magnetic force on an imaginary free N-pole is zero. No lines of force can pass through a neutral point.

Field due to a bar magnet

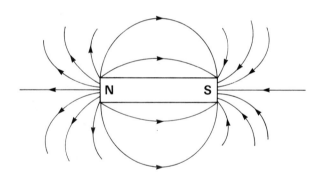

Figure 8.7

Field due to unlike and like poles

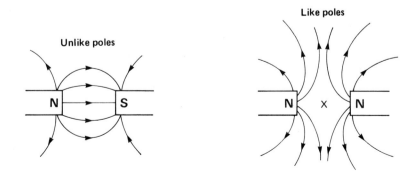

Figure 8.8

Field due to parallel magnets

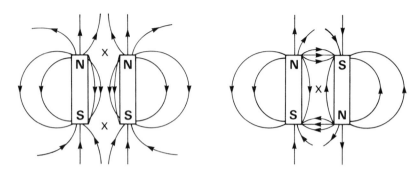

Figure 8.9

Earth's magnetic field

The Earth is like a huge bar magnet, with a magnetic pole almost coincident with the geographic North pole. See Fig. 8.10.

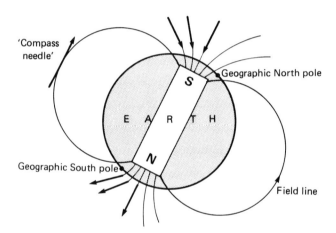

Figure 8.10

The arrows marked in Fig. 8.10 represent compass needles. So a compass needle at a point near the geographic North pole would be tilted almost straight downwards if *freely* suspended.

Inclination or dip

This is the angle between the direction of the Earth's magnetic field and the horizontal. Consequently it varies over the surface of the Earth.

Magnetic declination

The **geographic meridian** at a place is a plane containing the place and the Earth's axis of rotation.

The **magnetic meridian** at any place is a vertical plane through that place containing the magnetic axis of a freely suspended magnet.

The **magnetic declination** is the angle between the magnetic and geographic meridians.

Magnets in the Earth's field

See Fig. 8.11.

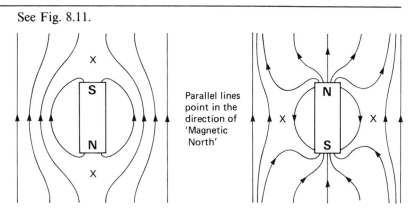

Parallel lines point in the direction of 'Magnetic North'

Figure 8.11

Magnetic effect of a current (Oersted's experiment)

The flow of electric charge in a wire produces a magnetic field which will deflect the needle of a compass if this is placed nearby. Reversing the current direction changes the needle deflection. See Fig. 8.12.

Steady current

Compass

Figure 8.12

Field due to a straight wire

Reversing the current *reverses* the direction of the field. See Fig. 8.13.

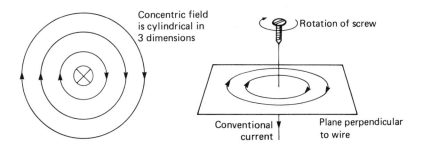

Concentric field is cylindrical in 3 dimensions

Rotation of screw

Conventional current

Plane perpendicular to wire

Figure 8.13

Maxwell's right hand screw rule

If a right hand screw moves with the conventional current, the direction of its rotation gives the field direction. See Fig. 8.13.

Field due to a circular coil

At the centre of the coil the field lines are straight and at right angles to the coil plane. See Fig. 8.14.

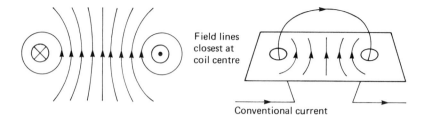

Figure 8.14

Field due to a solenoid

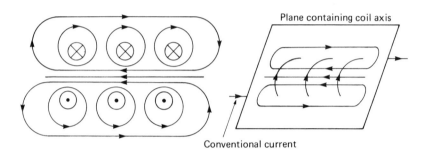

Figure 8.15

The field inside the solenoid can be made very strong if it has a large number of turns or a larger current passes. Its direction may be determined by the right hand grip rule. See Fig. 8.15.

Electromagnet

This is a coil of wire round a soft iron core; see Fig. 8.16. The strength of an electromagnet increases if (a) the current increases, (b) the number of turns increases, (c) the poles are closer together. The electromagnet is temporary. In the horse-shoe shaped electromagnet the coil must be wound in opposite directions on each prong of the horse-shoe to comply with the right hand grip rule.

Applications of the electromagnet include the electric bell, relay, lifting scrap metal, telephone earpieces, electric motors and dynamos.

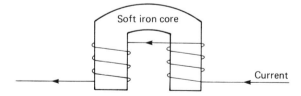

Figure 8.16

Force on a current carrying conductor in a magnetic field

A current carrying conductor in a magnetic field experiences a force and if the conductor can move, it does. It experiences *no* force if it is *parallel* to the field.

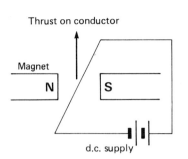

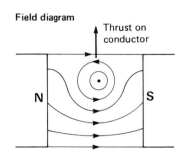

Figure 8.17

This is called the **motor effect** and if the direction of the current or of the field is reversed, the conductor will move in the opposite direction. From the field diagram in Fig. 8.17, it can be seen that the fields of the magnet and the conductor oppose each other (effectively 'cancelling') at the top but have the same direction at the bottom. Since they repel each other sideways there is an overall *push upwards*.

Fleming's left hand rule

See Fig. 8.18.

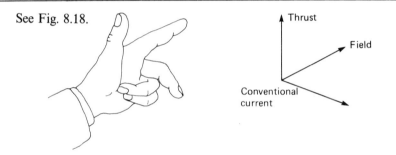

Figure 8.18

Effect of a magnetic field on a stream of charges

Conventional current has the opposite direction to electron flow. See Fig. 8.19.

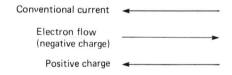

Figure 8.19

From noting these directions the force on a stream of charged particles can be found, bearing in mind that Fleming's left hand rule considers conventional current. See Fig. 8.20.

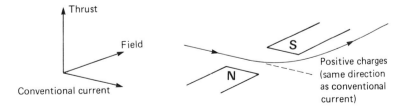

Figure 8.20

Simple d.c. motor

Electric motors are used in vacuum cleaners, washing machines, mowers, lifts, video recorders, etc. Windscreen wipers are driven by one and the engine is started by another. The motor is made up of a rectangular coil which can rotate between the poles of a magnet. Each end of the coil is connected to a power supply via a commutator.

Factors affecting the force on a current carrying conductor in a magnetic field

The factors affecting the force on a current carrying conductor *perpendicular* to a magnetic field can be shown by the use of a simple current balance set up as shown in Fig. 8.21 and working off a smoothed d.c. supply.

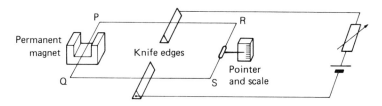

Figure 8.21

The pointer is attached to a *piece of insulator* at the middle of RS so current only passes through PQ. The frame is *initially balanced* on the knife edges but becomes *tilted* when current is passed. Metal 'riders' can be placed along the frame to *restore balance*, and if the current is increased more of them are required.

By putting more magnets side-by-side along PQ the length of conductor in the magnetic field is increased and once again more riders are needed to restore balance.

Finally if the permanent magnet is replaced by a *coil* of many turns through which there is a current passed from a separate source, the strength of the field acting on the conductor *can be altered*. Stronger magnets also produce an increase in the force acting on the conductor. The position of the coil can be changed, so the direction of the field with respect to the conductor is altered.

Magnetic flux density, *B*

The force F on a conductor length L and carrying current I, which is *perpendicular* to a uniform magnetic field, is directly proportional to those three quantities, and is perpendicular to the plane containing the field and the conductor.

The **magnetic flux density** or **magnetic induction** B is the force that acts per unit length upon a conductor carrying unit current lying perpendicular to the magnetic field.

$$B = \frac{F}{IL}$$ (units: $N\,A^{-1}\,m^{-1}$)

(The units $N\,A^{-1}\,m^{-1}$ are called the tesla and have the symbol T.)

The magnetic flux density is a vector quantity whose magnitude can be expressed (apart from the above equation) as the number of field lines passing through unit area. The direction of B is given by the *direction* of the *field line* at a point.

Rearranging the equation for B,

$$F = BIL$$

However, if the conductor is set at an angle θ to the field (Fig. 8.22):

$$F = BIL \sin \theta$$

Therefore when $\theta = 90°$ then F has its *maximum value*.

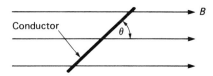

Figure 8.22

Examples of magnetic flux density

Example	Magnetic flux density (T)
Magnets on steel yoke	2.5×10^{-2}
Earth's magnetic field in Britain	5.3×10^{-3}
Fairly strong magnet	1

Comparison of magnetic flux density

The magnetic flux density produced by a conductor can be investigated using a *search coil*. This is a compact coil of about 5000 turns connected to an *oscilloscope*.

This indicates the *emf* induced in the coil by the changing field around the conductor which acts along the axis of the coil. There are two ways in which this changing field is obtained.

Firstly, by using a.c. in the conductor, as for example when investigating the magnetic flux density near a straight conducting wire. See Fig. 8.23.

Secondly, a changing magnetic field on d.c. in the conductor only occurs immediately on switching on or off the current. The search coil in this case is connected to a *light beam galvanometer* which registers a quick deflection on switching on or off the current in the conductor, as for example when investigating the variation of magnetic flux density in a solenoid. See Fig. 8.24.

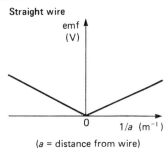

Figure 8.23

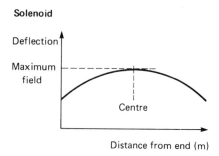

Figure 8.24

For a long, straight solenoid

$$B = \frac{\mu_0 N I}{L}$$

B is the magnetic flux density at a point on the solenoid axis near the *solenoid centre*. N is the number of turns on the coil. L is the length of the solenoid and I is the current passing through it.

μ_0 is called the **permeability of free space** or the **magnetic constant** and has the value

$$\mu_0 = 4\pi \times 10^{-7}\,\mathrm{H\,m^{-1}}$$

(H is the symbol for the henry, the unit of electric inductance.)

Sometimes B is written as $B = \mu_0 N^* I$, where N^* is the number of turns per unit length.

For a point at the end of a long solenoid

$$B = \mu_0 N I / 2L$$

For a long, straight wire

$$B = \frac{\mu_0 I}{2\pi a}$$

where a is the distance from the wire.

This is the magnetic flux density at a point of *perpendicular distance a* from the wire carrying current I.

The equations above apply to *ideal, infinitely long* conductors but are very nearly true for the conductors investigated in the laboratory.

Forces on charges in a magnetic field

For a current carrying conductor *perpendicular* to a uniform magnetic field, the force is

$$F = BIL$$

where L is the length of the conductor in the field. Now

$$I = \frac{Q}{t} \quad \text{and} \quad F = \frac{BQL}{t}$$

where Q is the charge passing through the conductor.

If the charge Q has a *drift velocity* v then $v = L/t$. Then

$$\boxed{F = BQv}$$

The force on each unit of charge e is

$$\boxed{F = Bev}$$

and this is only true if the charge is travelling *perpendicularly* to the field with a drift velocity v and applies to charges moving outside of conductors too.

The direction of the force is given by *Fleming's left hand rule*, remembering the second finger indicates the direction of *conventional current* or positive charge.

When charged particles pass into a *magnetic field* perpendicular to it, they are deflected into a *circular path*. If $\theta \neq 90°$ the path is a *helix. No force* exists when $\theta = 0°$.

Helices are characteristic tracks for electrons in bubble-chamber films taken in particle physics experiments.

Ampère

This is defined in terms of the force between two infinitely long, parallel wires which arises because of the *magnetic fields* produced by them (see Fig. 8.25).

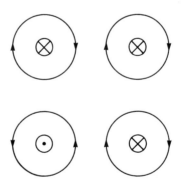

Figure 8.25

Currents going in opposite directions give rise to repulsion.

Consider two wires X, Y carrying currents I_x, I_y, as shown in Fig. 8.26.

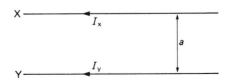

Figure 8.26

The magnetic flux density at Y due to the current I_x is

$$B_x = \frac{\mu_0 I_x}{2\pi a}$$

Therefore the force on length L of Y is

$$F = B_x I_y L$$

so

$$\boxed{F = \frac{\mu_0 I_x I_y L}{2\pi a}}$$

(An *equal* but *opposite* force acts on wire X.)

For the following, assume that the permeability of a vacuum is

$$\mu_0 = 4\pi \times 10^{-7}\,\mathrm{H\,m^{-1}}$$

Example 8.1

A current of 6.4 A passes along a straight conductor which is located in a uniform magnetic field of flux density 4.3×10^{-4} T. If the wire is 4.8 m long, determine the force upon it due to the field if it lies at (i) 90° and (ii) 24° with respect to the field lines.

(i)   The force on the wire is

$$F = BIL = 4.3 \times 10^{-4} \times 6.4 \times 4.8 = \underline{0.013\,\mathrm{N}}$$

(ii) If the wire is at an angle θ with respect to the field, then

$$F = BIL\sin\theta = 4.3 \times 10^{-4} \times 6.4 \times 4.8 \times \sin 24° = \underline{5.4 \times 10^{-3}\,\mathrm{N}}$$

Example 8.2

The magnetic field which results when 3.4 A of current passes along a vertical straight conductor and the horizontal component of the Earth's magnetic field B_h produce a neutral point, at a distance of 0.04 m from the centre of the conductor perpendicular to its axis. Find the value of B_h.

The flux density at a distance a from the centre of the conductor is

$$B = \frac{\mu_0 I}{2\pi a} = B_h \qquad \text{(at the neutral point)}$$

Therefore

$$B_h = \frac{4\pi \times 10^{-7} \times 5}{2\pi \times 0.04} = \underline{1.7 \times 10^{-5}\,\mathrm{T}}$$

Example 8.3

An ion of charge Q is projected into a uniform magnetic field, in the same direction as the field but at an angle of θ with respect to the field lines. If the ion has an initial velocity v, describe the path it will take.

See Fig. 8.27. From Fleming's left hand rule, the motion of the ion, the

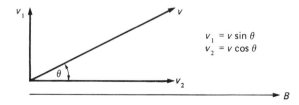

$$v_1 = v\sin\theta$$
$$v_2 = v\cos\theta$$

Figure 8.27

force upon it and the field are all perpendicular to each other. Thus, the force due to the magnetic field perpendicular to the field is

$$F = BQv \sin \theta = \frac{m(v \sin \theta)^2}{r}$$

where m is the mass of the ion.

The force parallel to the field is zero. Therefore

$$r = \frac{mv \sin \theta}{QB}$$

Thus, the ion is forced to travel in a vertically circular path which is extended in the direction of the field lines due to the component v_2 of the initial velocity v; i.e. the ion follows a 'corkscrew' path (Fig. 8.28).

Path of the ion

B

Figure 8.28

Example 8.4

From the definition of the ampère, find the value of μ_0, the permeability of free space.

The force between two wires carrying currents I_x and I_y separated by a distance a along a length L is

$$F = \frac{\mu_0 I_x I_y L}{2\pi a}$$

In this example, from the definition of the ampère,

$$I_x = I_y = 1\,\text{A}$$
$$F/L = 2 \times 10^{-7}\,\text{N m}^{-1}$$

and

$$a = 1\,\text{m}$$

Therefore

$$\mu_0 = \frac{2\pi a F}{I_x I_y L} = 4\pi \times 10^{-7}\,\text{H m}^{-1}$$

Force on a long, straight conductor in a solenoid

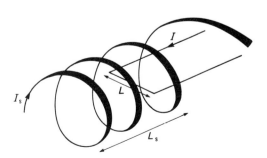

Figure 8.29

See Fig. 8.29. If the solenoid has N turns, the magnetic flux density at its centre along its axis is given by

$$B = \frac{\mu_0 N I_s}{L_s}$$

where I_s and L_s are the current in the solenoid and the length of the solenoid respectively. The force on the perpendicular wire L carrying a current I inside the solenoid at its centre is

$$F = BIL$$

and so

$$F = \frac{\mu_0 N I_s I L}{L_s}$$

Simple current balance

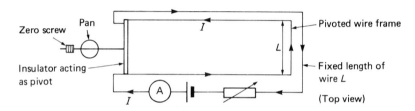

Figure 8.30

The simple current balance (Fig. 8.30) employs the relation for magnetic force between two wires carrying the same current.

Theory of the simple current balance

Before the current balance is applied to a circuit, the zero screw is first used to *balance the apparatus*. When current is switched on a force is produced between the two wires in the balance running *parallel* to each other and of length L. As the currents are in opposite directions, the balance is destroyed and can only be restored by *adding a mass* to the pan.

If the separation of these two wires is a the force which arises between them is

$$F = \frac{\mu_0 I^2 L}{2\pi a}$$

where I is the current in the circuit, to be measured.

At balance when a mass m has been added to the pan,

$$mg = \frac{\mu_0 I^2 L}{2\pi a}$$

where g is the acceleration due to gravity.

Rearranging the above equation to give current:

$$I = \sqrt{\frac{2\pi \, mag}{\mu_0 L}}$$

The current I will be in *ampères* if the other terms are in SI units.

A larger force is produced (and greater accuracy achieved) if the balance frame is inserted inside a long, flat solenoid carrying the current to be measured.

Rectangular current carrying coil in a magnetic field

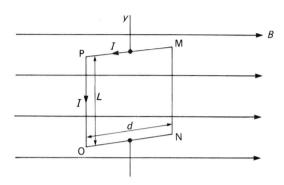

Figure 8.31

The *rectangular coil* (Fig. 8.31) has N turns and can rotate about the y-axis. Figure 8.32 shows the arrangement looking from the top.

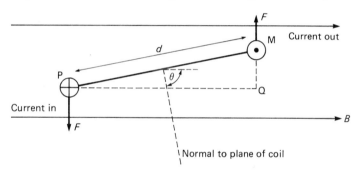

Figure 8.32

Rectangular coil theory

The vertical sides of the coil OP and NM (see Fig. 8.31) have forces F on them which are *equal*, but *opposite*:

$$F = BILN$$

The forces will remain constant as the coil rotates since OP and NM are always *perpendicular* to the uniform magnetic field of flux density B. These two forces form a *couple* whose *moment* or *torque* is

$$C = F \times \text{the perpendicular distance between the lines of action of the two forces}$$

$$= F \times \text{PQ} \quad \text{(in Fig. 8.32)}$$

$$= Fd \sin \theta$$

and so

$$C = BILNd \sin \theta$$

If the area of the coil is $A = dL$, then

$$C = BIAN \sin \theta$$

The *couple* causes the coil to rotate until $\theta = 0°$ when the moment of the couple $= 0$.

Electromagnetic moment

$$m = IAN$$

The quantity n is called the **electromagnetic moment** of the coil, and so the moment of the couple acting on the coil can be written as

$$C = Bm \sin \theta$$

Moving coil galvanometer

This measures *small currents* (e.g. mA) or *p.d.'s* (e.g. mV), although the scale is not usually *marked off* in either. At the heart of the instrument is a *rectangular coil* and this is mounted on jewelled bearings. Insulated copper wire makes up the coil. See Fig. 8.33.

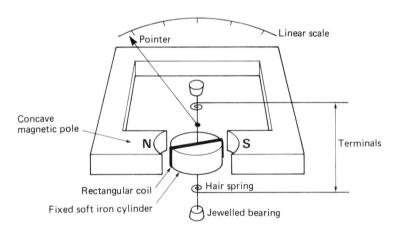

Figure 8.33

When current starts passing through it, it begins to rotate until stopped by hair springs which are attached to its mounting.

The greater the deflection of the pointer which is also attached to the mounting, the greater the current.

A *fixed* soft iron cylinder around which the coil rotates and the concave poles of the magnet providing the field in which the coil turns produce a radial field. All the field lines pass the *centre* of the cylinder.

This means that the divisions on the meter are the same size (i.e. a *linear scale*). The field lines are also always parallel to the plane of the coil; hence the deflecting moment C of the coil is always a *maximum* ($\theta = 90°$). Therefore

$$C = BIAN$$

where A is the mean area of the coil plane and N is the number of turns. B is the uniform magnetic flux density and is uniform only if the air gap is so too.

When C, the moment of the couple due to the coil, is equal to the moment of the couple due to the hair springs (but opposite in direction), *rotation stops*. Then

$$BIAN = k\alpha$$

where k is the moment of the couple due to the springs per unit deflection and α is the deflection.

The sensitivity of the galvanometer is improved by:
(1) more turns on the coil,
(2) a stronger magnet,
(3) weaker hair springs, longer pointer or mirror attached to the coil (light beam galvanometer).

Measuring larger currents

The meter only passes small currents, and to measure larger ones a resistor called a *shunt* is placed *in parallel* with it. The shunt diverts most of the current *past* the meter (Fig. 8.34).

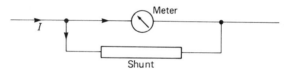

Figure 8.34

For example a meter may have a *full-scale deflection* (fsd) of only 2 mA and a resistance of 10 Ω. If it needs to read 2 A at fsd, then 1.998 A must pass *via* the shunt of resistance S.

Now,

$$\text{p.d. across galvanometer} = \text{p.d. across the shunt}$$
$$0.002 \times 10 = 1.998 \times S$$

The value of S is

$$S = \frac{0.002 \times 10}{1.998} = \underline{0.01\ \Omega}$$

Shunts are made of thick manganin wire, whose resistance changes little with temperature. Ammeters are used *in series* in circuits. So as not to alter current readings, they have a *low* resistance. For the example above, the total resistance of the meter R is given by

$$1/R = 1/10 + 1/0.01$$
$$\text{i.e. } R = \underline{0.01\ \Omega}$$

Measuring larger p.d.'s

A resistor of high value called a **multiplier** is put in series with the meter (Fig. 8.35). If the fsd needed is 2 V, then

$$\text{p.d. across multiplier} + \text{p.d. across galvanometer} = 2\ \text{V}$$
$$0.002 \times M + 0.002 \times 10 \qquad\qquad = 2\ \text{V}$$

Thus,
$$M = \underline{990\,\Omega}$$

Voltmeters are connected *in parallel* and have *high* resistances.

Figure 8.35

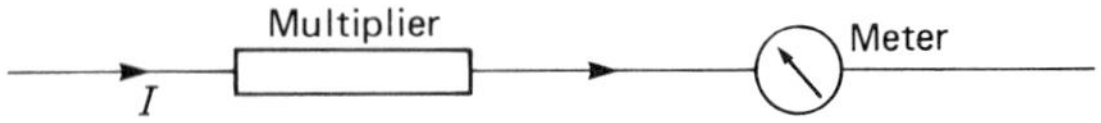

Sensitivity of the galvanometer

At deflection, the moment of the couple due to the coil is equal to the moment of the couple due to the hair springs. That is,

$$\boxed{BIAN = k\alpha}$$

where k is the moment of the couple due to the hair springs per unit deflection and α is the deflection. k has units of N m rad^{-1} and α is in radians.

From the above equation, the current is directly proportional to the deflection if the magnetic flux density is constant, and thus the current sensitivity of the meter is given by

$$\boxed{\alpha/I = BAN/k}$$

Therefore to *increase* the sensitivity of the galvanometer:

(1) B can be increased by making the magnet stronger or the air gap between the magnet and the coil smaller,
(2) A can be increased by enlarging the coil, although this may cause oscillations about the reading on the scale,
(3) N can be increased,
(4) k can be increased, although if the springs are made weaker, time for reading is increased.

The expression for p.d. sensitivity is:

$$\boxed{\alpha/V = BAN/kR}$$

where R is the resistance of the coil.

Galvanometers work best when their *resistance matches* the resistance of the circuit they are connected to.

Circuit resistance	Galvanometer resistance	Sensitivity
Low	Low	High α/V
High	High	High α/I

To prevent oscillations of a pointer about a reading, the coil is *critically* damped. Greater accuracy is obtained by attaching a mirror above the coil on the mounting. A large deflection of light reflected from this mirror can then be observed on a scale when a small rotation of the coil occurs.

Electromagnetic induction

This is the production of electricity from magnetism. Two methods of induction are now considered.

Method 1

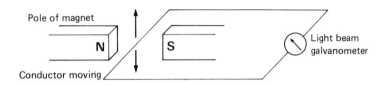

Figure 8.36

The deflection produced on the galvanometer (Fig. 8.36) is in *opposite* directions when the conductor moves up or down between the poles of the magnet, and it only lasts so long as the conductor is *moving*.

Method 2

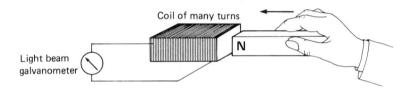

Figure 8.37

As the magnet is *moved*, a current is induced in the coil (Fig. 8.37). The deflections produced are in opposite directions on entry and withdrawal. They again last only as the magnet is moving. An *emf* is induced across a conductor when it cuts magnetic field lines, i.e. moves across them but not along them, or if the number of magnetic field lines passing through the conductor (*linking it*) changes.

If the conductor is part of a circuit, an induced current is also obtained. The induced emf *increases* as:

(1) the speed of the magnet in the coil increases,
(2) the number of turns on the coil increases,
(3) the strength of the magnet increases.

Instead of using a permanent magnet, a current carrying solenoid may be employed, which produces a similar magnetic field. Also, instead of moving the solenoid, a changing current is used.

Current changes very rapidly when it is first switched on and when it is switched off. However, rather than using a switch all the time, a continuously changing current is supplied in the form of *a.c.*

Mutual induction

This is the formation of an induced emf in one circuit as the result of current changes in another *separate* circuit.

It is upon this principle that the transformer (and the search coil) is based.

The coil through which the current is initially passed is called the **primary**, and that in which an emf is induced is the **secondary**. Placing a soft iron rod in the coils increases the emf. It is further increased if the coils are wound onto an iron ring as this concentrates the magnetic field lines.

If the secondary forms part of a circuit, a.c. in the primary will give a.c. in the secondary.

Magnetic flux

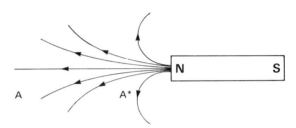

Figure 8.38

$$\Phi = B \times A$$

where Φ is called the magnetic flux.

If a coil is moved from A to A* (say) in Fig. 8.38, the number of magnetic field lines linking it increases. The magnetic field density B can be represented as the number of field lines penetrating unit area. So the magnetic flux, which is the product of the cross-sectional area A of a coil and B, represents the number of field lines linking that coil.

Definition of magnetic flux

Magnetic flux is defined as the product of the flux density normal to a small plane surface and the surface area A.

The unit of magnetic flux is the **weber** (symbol Wb).

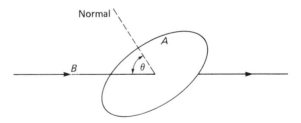

Figure 8.39

See Fig. 8.39. If the magnetic flux density B is at an *angle* θ to the normal to the surface, the magnetic flux is

$$\Phi = BA \cos \theta$$

Flux linkage

For a coil of cross-sectional area A and N turns, the magnetic flux $N\Phi$ is called the **flux linkage** since the same magnetic flux links all the turns.

Faraday's law of electromagnetic induction

This states that the induced emf ε is directly proportional to the rate of change of flux linkage $N\Phi$ or the rate of flux cutting.

$$\varepsilon \propto \frac{\mathrm{d}}{\mathrm{d}t}(N\Phi)$$

For SI units,

$$\varepsilon = \frac{\mathrm{d}}{\mathrm{d}t}(N\Phi)$$

The formal definition of the **weber** can now be given.

The *magnetic flux* which induces in a one turn coil an emf of 1 V when the flux is reduced to zero in 1 s is one weber.

Example 8.5

A coil of 10 turns and cross-sectional area 5 cm² is at right angles to a flux density of 2×10^{-2} T which is reduced to zero in 10 s. Find the flux change and the induced emf.

The *flux linkage* has been reduced from a value $NBA = 10 \times 2 \times 10^{-6} \times 5$ to zero and therefore the change is

$$1.0 \times 10^{-4} \text{ Wb}$$

The change in time is 10 s, and so the *induced emf* is

$$\varepsilon = \frac{1.0 \times 10^{-4}}{10} = 1.0 \times 10^{-5} \text{ V}$$

Faraday's law can be demonstrated by using a primary of many turns connected to a *signal generator*. A ten-turn secondary connected to an oscilloscope is wound onto the primary. As the frequency of a.c. in the primary is increased, the rate of *flux linkage increases* and the *emf* shown on the *oscilloscope also increases*.

(Note that although only flux linking is demonstrated, flux linking and flux cutting are *equivalent*.)

Lenz's law

This states that the direction of the induced emf is such that it opposes the flux change producing it.

When a permanent magnet is pushed into, or withdrawn from, a coil connected to a galvanometer, for example, the deflections observed are in opposing directions. This is because the direction of the induced current (*created by* the induced emf) produces a magnetic field which prevents the entry of the magnet if it is being pushed in, or prevents its exit if it is being withdrawn. See Figs 8.40 and 8.41.

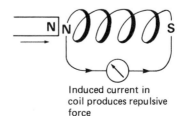

Induced current in
coil produces repulsive
force

Figure 8.40

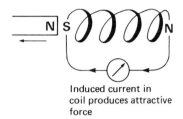

Induced current in
coil produces attractive
force

Figure 8.41

The *conservation of energy* is also shown here since work has to be done against the magnetic force produced by the induced current. That is, mechanical energy is converted into electrical energy—and so the induced current is not free!

As a result of Lenz's law a *negative sign* is introduced into the expression for induced emf:

$$\varepsilon = -\frac{\mathrm{d}(N\Phi)}{\mathrm{d}t}$$

Fleming's right hand rule (or dynamo rule)

This is the second of Fleming's rules and gives the direction of the induced current (conventional), produced in a conductor which is moving in a perpendicular direction, relative to a magnetic field. See Fig. 8.42.

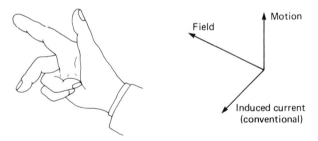

Field

Motion

Induced current
(conventional)

Figure 8.42

Eddy currents

When conductors move in a magnetic field or when the magnetic field changes around them, electric currents are induced in them. These are called **eddy currents** and can be substantial because of the low resistance of the paths they take. They have two effects—magnetic and heating. An example of the former is the critical damping produced by the soft iron cylinder in a moving coil galvanometer. An example of the latter is the refinement of germanium for use in semiconductor devices.

Induced emf for a rotating coil

A coil of N turns with cross-sectional area A rotates perpendicularly relative to a uniform magnetic field of flux density B (Fig. 8.43). The flux linking every turn is

$$\Phi = BA\cos\theta$$

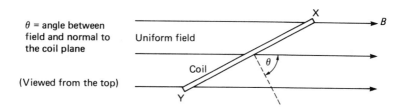

θ = angle between field and normal to the coil plane

Uniform field

Coil

(Viewed from the top)

Figure 8.43

If ω is the angular velocity,

$$\Phi = BA \cos \omega t \qquad (\text{since } \theta = \omega t)$$

From Faraday's law of *electromagnetic induction*,

$$\varepsilon = -\frac{d}{dt}(N\Phi) = -\frac{d}{dt}(NBA \cos \omega t)$$

Now B, N and A are constants and can be taken out of the brackets:

$$\varepsilon = -BAN \frac{d}{dt}(\cos \omega t)$$

Finally, performing the differentiation,

$$\boxed{\varepsilon = BAN\omega \sin \omega t}$$

Generators

In a simple generator, a rectangular coil is rotated in a uniform magnetic field between the poles of a permanent magnet. If a.c. is to be produced, slip rings are required at the ends of the coil to take away the current via brushes.

For the coil (see Fig. 8.44),

$$\Phi = BA \cos \omega t \quad \text{and} \quad \varepsilon = BAN\omega \sin \omega t$$

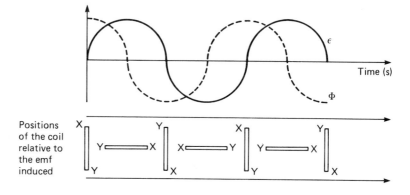

Figure 8.44

D.c. generators use a split copper ring called a **commutator** to take off the current and this ensures that, although the induced emf is changing in magnitude, its direction is always *the same*.

In real generators, an armature is used where several coils are wound onto a soft iron core. The armature is laminated to reduce eddy currents.

Example 8.6

A rectangular coil of 250 turns is supported by two hair springs which hold it in a vertical position in a uniform magnetic field of flux density 8.6×10^{-4} T. The plane of the coil has area $16 \, \text{cm}^2$ and the normal to the plane is at an angle of $\theta = 35°$ to the horizontal magnetic field lines. If the current passing into the coil is 5 A, determine (i) the moment of the couple due to the hair springs preventing further rotation of the coil, and (ii) the new value of the angle θ which would be produced by doubling the area of the plane of the coil.

(i)   See Fig. 8.45.

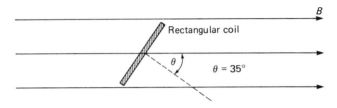

Figure 8.45

The moment of the couple due to the hair springs is given by

$$C = BIAN \sin \theta$$

Therefore

$$C = 8.6 \times 10^{-4} \times 5 \times 16 \times 10^{-4} \times 250 \times \sin 35°$$
$$= \underline{1 \times 10^{-3} \, \text{N m}}$$

(ii)   Now

$$C = BIAN \sin \theta$$

Rearranging

$$\sin \theta = C/BIAN$$

Thus,

$$\sin \theta = (1 \times 10^{-3})/(8.6 \times 10^{-4} \times 5 \times 32 \times 10^{-4} \times 250)$$

Therefore

$$\theta = \underline{16.9°}$$

Example 8.7

The plane of a 900 turn rectangular coil has area $20 \, \text{cm}^2$. The coil is supported on a vertical axle about which it can rotate, and lies in a horizontal magnetic field of flux density 5×10^{-2} T. If it carries a current of 4.5 A, find (i) the maximum induced emf, and if it rotates 30 times per second, find (ii) its maximum power output.

(i)   The emf induced across the coil is given by

$$\varepsilon = BAN\omega \sin \omega t$$

but

$$\omega t = \theta \quad \text{and} \quad \omega = 2\pi f$$

Thus, the maximum induced emf will be given by

$$\varepsilon = BAN\omega \sin 90°$$
$$= 5 \times 10^{-2} \times 20 \times 10^{-4} \times 900 \times 2\pi \times 30 = \underline{17 \, \text{V}}$$

(ii)   The power output from the rotating coil is

$$P = \varepsilon I = 17 \times 4.5 = \underline{76.5 \text{ W}}$$

Example 8.8

If the coil of example 8.7 forms part of a motor and has a resistance of 68 Ω, find (i) the motor's back emf, and (ii) the supply p.d. across which the motor is connected.

(i)   The back emf is the induced emf across the coil, i.e. $\underline{17 \text{ V}}$.

(ii)   If the supply p.d. is V, then

$$V - \varepsilon = IR$$

Therefore

$$V = IR + \varepsilon = (4.5 \times 68) + 17 = \underline{323 \text{ V}}$$

Simple a.c. generator

This consists of a rectangular coil, which is able to rotate between the poles of a permanent magnet. As the coil rotates an *emf is induced* across its ends which in turn provides a current to an external circuit. Each end of the coil is attached to a *slip ring* which remains in contact with one carbon 'brush' as it spins round with the coil. The carbon brushes allow the current to pass into the external circuit. See Fig. 8.46.

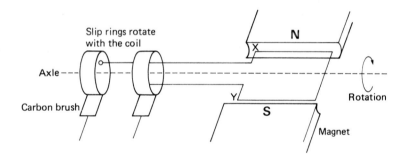

Figure 8.46

The direction of the induced emf changes every time the coil is rotated through 180° and because each slip ring is in contact with the *same* carbon brush all the time, the direction of the current in the external circuit also changes (see Fig. 8.47).

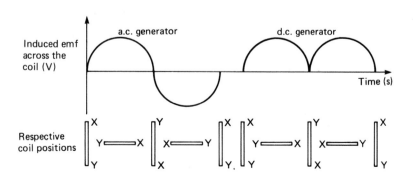

Figure 8.47

In effect this means that the electrons move continuously, backwards and forwards throughout the circuit—not really getting anywhere. Mains supply is a.c., of frequency 50 Hz. This means that mains current is changing direction one hundred times each second.

Simple d.c. generator

When the slip rings are replaced with a *commutator* a direct current is produced. As the coil goes through the vertical position each part of the split ring touches the *other* carbon brush and, because of this, every time the emf changes direction each brush continues to take current only in *one* direction (see Figs 8.47 and 8.48).

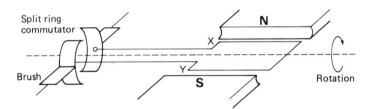

Figure 8.48

Back emf

The simple d.c. generator is much the same as a simple d.c. motor. A motor when it is working also produces an emf. It is called the **back emf**. It opposes and is almost equal to the applied p.d. across the motor. The current in the motor when it is going is therefore small or close to zero. However, the back emf takes a little time to build up and when the motor is first switched on the current through it is *large*. A high value resistance is applied to practical motors during this build up.

Practical generators

The emf produced by a generator or dynamo can be increased (i) by using a coil of more turns and (ii) by increasing the strength of the magnetic field—by employing an electromagnet and finally by making the coil rotate faster.

Transformers

When a magnet is pushed in and out of a coil of wire, an alternating emf is induced across the coil. If the moving magnet is itself replaced by a stationary coil of wire through which a current is switched on and off, an *alternating emf* is once again induced (see Fig. 8.49).

As the current is switched on it creates a magnetic field which builds up, acting like a moving magnet. When the current is switched off the magnetic field dies away, again behaving like a moving magnet.

A *changing current* (which produces a changing magnetic field) can be obtained by using a.c.

If both coils are wound onto a soft iron core, the effect is increased because the core concentrates the changing magnetic field around the

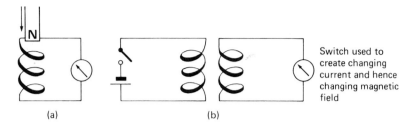

Figure 8.49   (a)   (b)   Switch used to create changing current and hence changing magnetic field

coils. The coil across which a.c. is applied is called the **primary** and the coil across which the emf is induced is called the **secondary**. See Fig. 8.50. Note that the transformer only operates on a.c. and the transformer effect is often referred to as **mutual induction**.

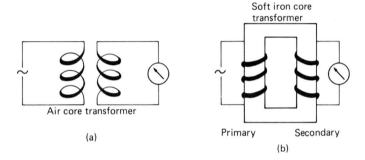

Air core transformer

(a)

Soft iron core transformer

Primary   Secondary

(b)

Figure 8.50

Transformer equation

Rather than switching on and off the current in the primary coil, a varying current can be obtained by simply *using a.c.* in the primary coil.

It is found from experiment that as the p.d. across the primary coil is *increased*, the induced emf across the secondary coil also *increases*. Conversely a decrease will show a decrease in the induced emf. If the number of turns on the primary coil *increases*, the induced emf across the secondary *decreases*, and vice versa.

These statements can be put in the form of an equation called the transformer equation:

$$\frac{V_s}{V_p} = \frac{N_s}{N_p}$$

where V_p and V_s stand for the applied p.d. across the primary and the induced emf in the secondary respectively, and N_p and N_s stand for the number of turns on the primary and on the secondary respectively.

Since the applied p.d. across the primary is an alternating one, the induced emf is also alternating. Thus, a transformer is a device which *changes* or *transforms* the *value* of an alternating voltage to another value.

Step-up transformer

A **step-up transformer** has more turns on the secondary than on the primary and so, from the equation, V_s is greater than V_p. See Fig. 8.51.

Step-down transformer

A **step-down transformer** has fewer turns on the secondary than on the primary and so, again from the equation, V_s is less than V_p. See Fig. 8.51.

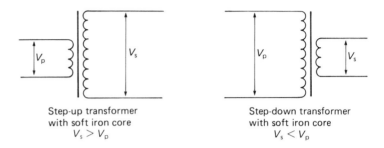

Step-up transformer
with soft iron core
$V_s > V_p$

Step-down transformer
with soft iron core
$V_s < V_p$

Figure 8.51

Energy losses

From the principle of conservation of energy, the electrical energy in the primary should equal the electrical energy in the secondary. Therefore

$$V_p I_p t = V_s I_s t$$

Rearranging gives

$$\boxed{\frac{V_p}{V_s} = \frac{I_s}{I_p}}$$

However, energy losses occur because of:

(1)   heating up of windings,
(2)   eddy currents (induced emf's in the *core* produce currents called eddy currents—these cause heating of the core and are reduced by laminating the core),
(3)   leakage of field lines—not all the magnetic lines of force from the primary cut the secondary coil.

Transmission of electricity

(1)   National grid—a network of cables connecting power stations to consumers. A high voltage is used to reduce energy loss.
(2)   Substations use transformers to lower voltage.

Reduction of energy losses

The losses that occur in the transformer are due to the heating of windings due to their resistance and are reduced by using thick copper wire. *Eddy currents* are reduced by *lamination*, and leakage of the field is cut down by the use of a complete soft iron core as opposed to, say, a C-core. Lastly, *hysterisis*—alternating magnetisation of the iron core causing heat—is lessened by employing a suitable magnetic material for the core.

Demonstration of the transformer equation

The equation relating the p.d.'s across the primary and secondary coils to the number of turns on the primary and the number of turns on the secondary is

$$\boxed{\dfrac{V_p}{V_s} = \dfrac{N_p}{N_s}}$$

This can be demonstrated by connecting the primary coil of a transformer to a *variable a.c. supply* and the secondary to an *oscilloscope*. If the ratio of turns is 1 : 1 (say both coils have 600 turns), the induced emf across the secondary shown on the oscilloscope should always match the applied p.d. across the primary. Next, if the p.d. across the primary is kept constant but the turns ratio varies, a corresponding change is observed in the induced emf.

Back emf in the transformer

The *alternating* magnetic flux due to the a.c. in the primary induces an emf in the secondary and also one *in itself*. This is called the **back emf** and it is almost *equal* to but *opposite* in direction to the applied p.d. on the primary. Thus, the net primary emf is *almost zero*.

Similar back emf's occur in dynamos and motors, resulting in small currents in the coils when the coils are actually *rotating*.

Self-induction

Any coil which induces in itself an emf due to the changing magnetic flux produced by a changing current through it undergoes **self-induction**. It *obeys* Faraday's law.

Changing current in coil

Changing magnetic flux linking the coil

Induced emf in the coil itself

The coil is called an **inductor**, with **inductance** (or **self-inductance**) and given the symbol L.

Circuit symbols

Air-cored inductor ⎯⌇⌇⌇⎯

Material-cored inductor ⎯⌇⌇⌇⎯

Demonstration with inductors: 1

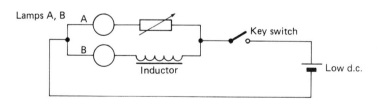

Figure 8.52

See Fig. 8.52. As current from the supply 'grows' on switching on, the induced emf in the coil *opposes* it. So a *delay* in the lighting of lamp B occurs.

Example 8.9

A step-down transformer has a primary coil of 450 turns which is applied across an a.c. supply of 240 V. Find the p.d. across the secondary if it has 330 turns and the transformer is 65% efficient.

Applying the transformer equation,

$$V_s/V_p = N_s/N_p$$

therefore

$$V_s = (V_p \times N_s)/N_p$$

Since the transformer is only 65% efficient

$$V_s = 0.65\,(V_p \times N_s)/N_p = 0.65\,(240 \times 330)/450$$

i.e.

$$V_s = \underline{114\ V}$$

Example 8.10

A generator produces electricity at a voltage of 15 kV, which is subsequently stepped up by a transformer to a value of 150 kV. This is then passed through cables of resistance 430 Ω to local power stations. If the power produced by the generator is 80 kW, find the ratio of the power loss in the cable without initially stepping up the voltage to the power loss when the voltage has been initially stepped up to 150 kV.

If the power is transmitted without stepping up the voltage

$$\text{cable current} = I = \frac{P}{V} = \frac{8 \times 10^4}{1.5 \times 10^4} = 5.3\ A$$

If the voltage is initially stepped up, and assuming 100% efficiency for the transformer, then

$$\text{cable current} = I' = \frac{P}{V'} = \frac{8 \times 10^4}{1.5 \times 10^5} = 0.53\ A$$

Thus, the ratio of power losses (let it be called k) is given by

$$k = I^2R/(I')^2R = 5.3^2/(0.53)^2 = \underline{100}$$

(The actual power losses are $I^2R = 5.3^2 \times 430 = 12.1\,kW$ and $(I')^2R = (0.53)^2 \times 430 = 121\ W$.)

Example 8.11

The electrical energy supplied to a transformer primary in 70 s is 150 kJ. If the transformer is only 69% efficient, calculate the power output of the secondary.

Since the transformer is only 69% efficient the amount of energy reaching the secondary in 70 s is

$$E_s = 0.69\,E_p = 0.69 \times 1.5 \times 10^5 = 1.04 \times 10^5\ J$$

Power is the rate of working, and therefore the power output of the secondary is given by

$$P_s = E_p/t = (1.04 \times 10^5)/70 = \underline{1.5\ kW}$$

(The power input for this transformer is given by $P_p = E_p/t = (1.50 \times 10^5)/70 = 2.1\,kW$.)

Demonstration with inductors: 2

See Fig. 8.53. On switching off, the *induced emf* in the inductor tries to oppose the *fall of current* and a brief flash occurs.

Sparking between switch contacts is another example of this effect.

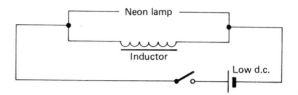

Figure 8.53

Inductance

The inductance L of a coil is defined as the ratio of the emf induced in it ε to the rate of change of current dI/dt to which it is due:

$$L = -\varepsilon \left/ \frac{dI}{dt} \right.$$

Since ε and dI/dt act in *opposition* to each other a negative sign has to be inserted.

The **henry** (symbol H) is the *inductance* of a circuit or coil in which an emf of 1 V is induced by a current change of 1 A per second.

$$1\,H = 1\,V\,s\,A^{-1}$$

Every circuit has *inductance* since every circuit has a field.

Energy stored in an inductor

Energy is stored by the inductor in the magnetic field it creates and is given by

$$\text{energy stored} = \tfrac{1}{2}LI^2$$

Mutual inductance

This is the process by which a changing current in a primary coil induces an emf in the secondary. The mutual inductance M for two coils is given by

$$M = -\varepsilon \left/ \frac{dI_p}{dt} \right.$$

where ε is the induced emf in the secondary and dI_p/dt is the rate of change of current in the primary coil.

The *henry* is then also the unit of *mutual inductance*.

Two coils have a mutual inductance of 1 H when an emf of 1 V is induced in the secondary for a primary current change of 1 A per second.

Capacitance

The ability to store charge, or **capacitance**, of an insulated conductor is the charge Q needed to produce a unit change in the potential of a conductor.

$$\boxed{C = Q/V}$$

The unit of capacitance is the **farad** (symbol F).

Earthing

The capacitance of the Earth is comparatively large and when conductors are connected to it, they become discharged and gain *zero potential*.

Action of a capacitor

A capacitor *stores charge* and is made up from two conductors or 'plates' separated by an insulator.

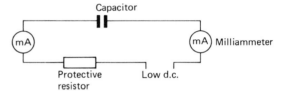

Figure 8.54

Equal deflections on the *milliammeters* (see Fig. 8.54) indicate an *equal charge* arrives at one plate to that which leaves the other. Both deflections are in the *same direction*. Since, when the capacitor is fully charged, the p.d. is equal but opposite to that across the d.c. supply the deflections are not permanent as electron flow then stops.

If the d.c. supply is *short circuited*, the capacitor acts as a cell and electrons flow back to the positive plate, once again giving a temporary deflection on each meter. The directions are opposite to what they were before.

Charge stored and factors affecting it

The charge stored by a capacitor is found from experiment to be given by

$$Q \propto \frac{VA}{d}$$

where A is the area of overlap of the plates and d is their separation. V is the applied p.d.

Since $Q = CV$,

$$C \propto \frac{A}{d}$$

Parallel plate capacitor

The constant of proportionality for the above equation is the permittivity, ε, so

$$C = \frac{A\varepsilon}{d}$$

The permittivity relates to the medium in which charges are situated and in the case of the capacitor relates to the material between the plates. This material is called a **dielectric**. The dielectric constant or **relative permittivity** ε_r is the ratio of the capacitance C of a capacitor with a *dielectric* to the capacitance C_0 with a *vacuum* between the plates.

$$\varepsilon_r = C/C_0 = \frac{\varepsilon A}{d} \bigg/ \frac{\varepsilon_0 A}{d}$$

Therefore

$$\varepsilon_r = \varepsilon/\varepsilon_0$$

So for a parallel plate capacitor

$$C = \frac{\varepsilon_0 \varepsilon_r A}{d}$$

The field produced by the charged plates extends *through* the dielectric and, as a result, charges develop either side of the dielectric opposite in sign to those on the plates. This causes a lowering of the p.d. across the plates and means more charge Q has to be supplied before the capacitor p.d. is equal to the applied p.d., causing a *rise in capacitance*.

Determination of relative permittivity

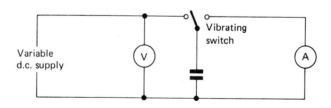

Figure 8.55

The *vibrating switch* (Fig. 8.55) charges and discharges the capacitor with a frequency f. Now

$$I = Q/t$$

and

$$Q = CV$$

Therefore

$$I = CV/t = CVf$$

The current I recorded on the ammeter is varied by changing the p.d. applied across the capacitor, V. A series of readings is then obtained and a graph of I versus V plotted. This will be a straight line whose *gradient* is Cf. If f is known, the value of C can be found. The capacitor used is one where the dielectric can be changed, and so the capacitance C for the dielectric under investigation is found and C_0 for air. The relative permittivity is then

$$\varepsilon_r = C/C_0$$

Electrometer/d.c. amplifier

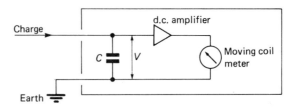

Figure 8.56   Earth

This instrument can be used to give a *direct reading* of charge Q arriving at the input, i.e. as a charge measuring device. See Fig. 8.56. A *known* capacitance C is applied across the input and a *moving coil meter* attached so that it will measure the p.d. V across it. This p.d. is small and has to be *amplified* before the meter will show a reading. This meter is suitably *calibrated*. A small *leakage current* occurs by the capacitor discharging through the input of the amplifier, but the rate of drop of p.d. is small. Then

$$\boxed{Q = CV}$$

The **electrometer** can be used to measure the capacitance C^* of a small capacitor. If this capacitor is charged up by a known p.d. V^* and placed across the input of the amplifier, it will *discharge*. Most of the charge Q from it then goes to the capacitor already across the amplifier and again

$$Q = CV \quad \text{and} \quad C^* = Q/V^*$$

To check that all the charge has been transferred, the electrometer capacitor C is *discharged* and the small capacitor C^* applied again across the input.

Capacitors in parallel

The charges are

$$Q_1 = C_1 V \quad \text{and} \quad Q_2 = C_2 V$$

Total charge is

$$Q = Q_1 + Q_2$$

Capacitance for capacitors in parallel

If the p.d. across two capacitors in parallel is V and the capacitances are C_1 and C_2 (see Fig. 8.57), the total charge is

$$Q = C_1 V + C_2 V = V(C_1 + C_2)$$

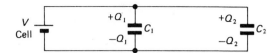

Figure 8.57

So the *combined capacitance* for capacitors *in parallel* is given by the sum of the individual capacitances:

$$C = C_1 + C_2$$

Generally

$$C = C_1 + C_2 + C_3 + \ldots$$

Also

$$\frac{Q_1}{Q_2} = \frac{C_1}{C_2}$$

Similarly for *more than* two capacitors in parallel.

Capacitors in series

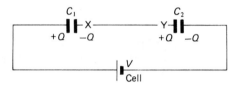

Figure 8.58

See Fig. 8.58. Electrons flow from the cell to capacitor C_2 causing a plate to acquire charge $-Q$ that *induces* a charge $+Q$ on the other plate. The electrons in connection XY go to X and induce *another* charge $+Q$ on the other plate of C_1. (A similar thing occurs for more than two capacitors in series.)

The p.d.'s of the capacitors are

$$V_1 = \frac{Q}{C_1} \quad \text{and} \quad V_2 = \frac{Q}{C_2}$$

The total p.d. is then

$$V = V_1 + V_2 = \frac{Q}{C_1} + \frac{Q}{C_2}$$

i.e. $\quad V = Q\left(\frac{1}{C_1} + \frac{1}{C_2}\right)$

Therefore the total capacitance C can be found from

$$\frac{1}{C} = \frac{1}{C_1} + \frac{1}{C_2}$$

Generally

$$\frac{1}{C} = \frac{1}{C_1} + \frac{1}{C_2} + \frac{1}{C_3} + \ldots$$

Energy of a charged capacitor

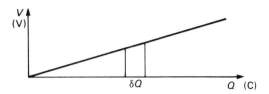

Figure 8.59

Since $Q = CV$ the graph of V versus Q (Fig. 8.59) is *linear*. If a capacitor discharges by a *small amount* δQ the small amount of *work done* δW is given by

$$\frac{\delta W}{\delta Q} = V$$

from the definition of p.d., where V is the p.d. across the discharging capacitor. Therefore $\delta W = V \delta Q$ which is the area of the strip shown in the graph in Fig. 8.59.

The *total work* done in discharging a capacitor, W, is given by the total area under the graph of V versus Q:

$$W = \tfrac{1}{2}QV$$

This is the *energy lost* by the capacitor since work W has been done in converting electrical energy stored in the electric field between the plates to *heat energy* in the discharging circuit. It is also equal to the *energy stored* by the capacitor and can be written as

$$\text{total energy} = \tfrac{1}{2}CV^2$$

R–C circuit discharging and charging characteristics

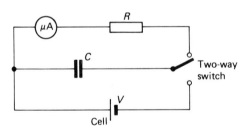

Figure 8.60

For the discharge characteristic curve the circuit in Fig. 8.60 is used. The capacitor is charged up by applying the 12 V p.d. across it, the contact between them then being broken. Discharge of the capacitor now takes place through the resistor and the values of current I are taken at 10 s intervals.

$$Q = CV = CIR$$

therefore

$$Q \propto I$$

The graph shows *exponential decay* and $t_{1/2}$ is shown as the **half-life** (Fig. 8.61).

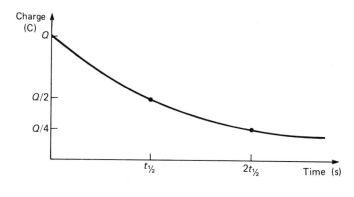

Figure 8.61

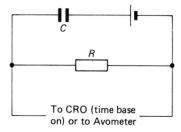

Figure 8.62

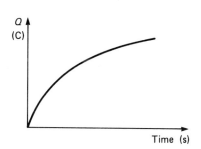

Figure 8.63

Both *charge* and *discharge* curves for a capacitor can be shown on an oscilloscope and charging is *also* found to be *exponential* (Figs 8.62 and 8.63).

The connecting wires in an *R–C circuit* have a small *self-inductance*, since all conductors produce *magnetic fields*. Initially during discharge the self-inductance *opposes* the discharge (Lenz's law), and when the capacitor is discharged it tries to keep the current going by charging up the plates in the reverse sense, and the current decreases until the p.d. is steady and the capacitor begins to discharge, and so on . . . .

For the following assume the permittivity of free space is

$$\varepsilon_0 = 8.9 \times 10^{-12} \text{ F m}^{-1}$$

Example 8.12

A school laboratory parallel plate capacitor is constructed by using two square metal plates of side 22 cm. These are set up vertically, 4 mm apart, and charged up using a d.c. 250 V supply. Determine the charge produced on the plates when (i) air and (ii) a dielectric of relative permittivity 6.8 separate the plates.

For a parallel plate capacitor,

$$Q = CV = \varepsilon_0 \varepsilon_r AV/d$$

where A is the area of overlap of the plates and d is their separation.

(i)   For air as the dielectric,

$$Q = \frac{8.9 \times 10^{-12} \times 1 \times 0.22^2 \times 250}{4 \times 10^{-3}} = \underline{2.7 \times 10^{-8}\ C}$$

(ii)   For the material of relative permittivity 6.8,

$$Q' = \frac{6.8 \times 8.9 \times 10^{-12} \times 0.22^2 \times 250}{4 \times 10^{-3}} = \underline{1.83 \times 10^{-7}\ C}$$

Example 8.13

A waxed paper capacitor consists of wax paper 10 cm wide keeping two strips of aluminium foil 0.06 mm apart. If the relative permittivity of the wax paper is 9.4 and the capacitance of the device is 1 μF find the length of the wax paper dielectric.

For the capacitor,

$$C = \varepsilon_0 \varepsilon_r A/d$$

Then the length of the dielectric

$$L = Cd/(\varepsilon_0 \varepsilon_r w)$$

where w is the width of the dielectric. Therefore

$$L = \frac{1 \times 10^{-6} \times 6 \times 10^{-5}}{8.9 \times 10^{-12} \times 9.4 \times 1 \times 10^{-1}} = \underline{7.2\ m}$$

Example 8.14

Identical capacitors of capacitance 2.5 μF are connected to a 3 V battery as shown in Fig. 8.64 (a) and (b). Determine the charge present on each capacitor in both circuits.

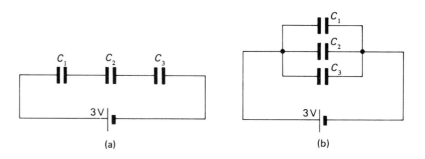

Figure 8.64   (a)   (b)

Circuit (a)—when the capacitors are in series, the total p.d. is the sum of the p.d.'s across the individual capacitors. Thus,

$$V = \frac{Q}{C_1} + \frac{Q}{C_2} + \frac{Q}{C_3}$$

therefore

$$Q = \frac{V}{(1/C_1 + 1/C_2 + 1/C_3)}$$

Then
$$Q = \underline{2.5 \times 10^{-6}\ \text{C}}$$

Circuit (b)—when the capacitors are in parallel
$$Q = CV = 2.5 \times 10^{-6} \times 3 = \underline{7.5 \times 10^{-6}\ \text{C}}$$

The repetition of discharge in an *R–C circuit* has the appearance of an *oscillatory* discharge which *dies* away. Most of the energy stored in the capacitor is converted to heat, but some is converted to electromagnetic radiation (*radio waves*). This results in interference on radio sets near discharging capacitors.

Using large resistances, however, gives damping of the oscillatory discharge and eventually critical damping where no oscillations occur, as in the above examples.

Alternating currents

An *alternating emf* produces an alternating current. The simplest form of the alternation is *sinusoidal*.

$$\boxed{\varepsilon = \varepsilon_0 \sin \omega t \quad \text{and} \quad i = i_0 \sin \omega t}$$

where $\omega = 2\pi f$, ε is the alternating emf and i is the alternating current. (The frequency f for the mains is 50 Hz.)

The *constants* ε_0 and i_0 are the amplitudes of the sinusoidal variations in emf and current respectively.

In a d.c. circuit, charge flows from one point to another, i.e. the current has the *same direction*. In an a.c. circuit, the charge moves forwards and backwards as the direction of the *current changes*.

Bridge rectifier a.c. meters

A *diode* is a component which allows current to pass only in *one direction*. The circuit symbol for a diode is ⟶(▶)⟶ where the arrow shows the direction of current allowed.

If the diode is connected to an a.c. supply only half the current passes through (Fig. 8.65).

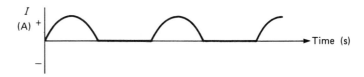

Figure 8.65

This is now *varying* but *direct current* and the process is called **half-wave rectification**.

The arrangement in Fig. 8.66 is called a **bridge rectifier**.

The diodes X and Y allow the current to pass in *one direction*, while L and M allow it to pass *only* when it *reverses*. So the moving coil meter at the centre always has current passing through it in the *same direction* (Fig. 8.67).

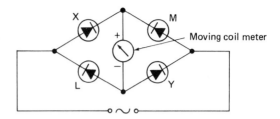

Figure 8.66

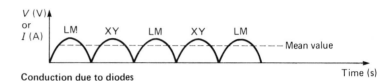

Figure 8.67   Conduction due to diodes

This is called **full-wave rectification**. The deflection on the meter is proportional to the mean value of the current and the scale is calibrated to read rms values for p.d. and current of sinusoidal form.

rms value of current

rms stands for root mean square and when applied to current it has the following definition: it is that steady direct current I which does work at the same rate as the alternating current i for a given resistance R. Thus,

$$I^2 R = (\text{mean value of } i^2)R$$
$$I^2 \quad = \text{mean value of } i^2$$

and so

$$I \quad = i_{\text{rms}}$$

If the a.c. is varying *sinusoidally* and has *peak value* i_0, then

$$i_{\text{rms}} = \sqrt{\text{mean value of } i_0^2 \sin^2 \omega t}$$
$$= i_0 \sqrt{\text{mean value of } \sin^2 \omega t}$$

From a graph of $y = \sin^2 \omega t$ (Fig. 8.68) it can be seen that the mean value is $1/2$.

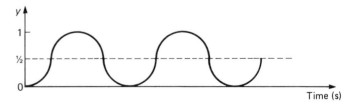

Figure 8.68

Thus,

$$\boxed{i_{\text{rms}} = \sqrt{1/2}\, i_0}$$

Similar expressions arise for the rms values of alternating p.d. and emf.

Experimental determination of rms value

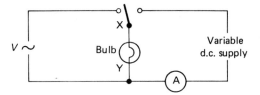

Figure 8.69

The switch at X in Fig. 8.69 is moved until adjustment of the variable d.c. supply lights the bulb to the same brightness as the a.c. supply. The rms value of the a.c. is then indicated by the reading on the d.c. ammeter. An oscilloscope connected across XY can act as a voltmeter. When set to d.c. it indicates the rms p.d., and on a.c. with zero time base it shows twice the peak current value.

The *peak value* for the mains p.d. is

$$v_0 = v_{rms} \bigg/ \sqrt{1/2} = \underline{339\ \text{V}}$$

Capacitors and a.c.

When a capacitor and a bulb are connected in series across a d.c. supply the bulb *does not* light. With an a.c. supply it does. The a.c. *appears* to pass through the capacitor because in one cycle it charges, discharges, charges in the opposite direction and discharges again the capacitor.

If the capacitance is reduced, the bulb gets dimmer, i.e. the a.c. is *opposed more* by *smaller capacitance*.

If the frequency of the a.c. is decreased, the same charge moves around the capacitor circuit in a longer time and so the current in the circuit is reduced.

Phase

The idea of **phase** can be applied to a.c. when considering the relationship between current and p.d.

Phase and a.c.

For an a.c. circuit with *only a resistor*, the current and the p.d. are *in phase* (Fig. 8.70).

For an a.c. circuit with *only a capacitor*, the current *leads* the p.d. by $\pi/2$ (Fig. 8.71).

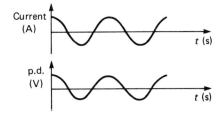

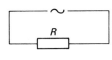

Figure 8.70

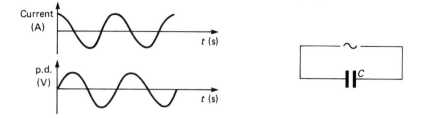

Figure 8.71

When the p.d. is *zero*, there is nothing to *oppose* charge movement so the current is a *maximum*. However, as the capacitor charges up, the p.d. across it rises until it reaches a maximum value (fully charged) and charge movement stops, i.e. zero current.

The capacitor begins to discharge and the current starts to increase, reaching a maximum value when the p.d. is zero. Now the capacitor begins to charge up in the opposite direction, and so on . . . .

$$i = \frac{dQ}{dt} = \frac{d}{dt} \; (Cv_0 \sin \omega t)$$

So

$$\boxed{i = \omega Cv_0 \cos \omega t}$$

Therefore the amplitude of the current is ωCv_0 when $\omega = 2\pi f$.

For an *inductor only* in an a.c. circuit, the current, p.d. and back emf are related as shown in Fig. 8.72.

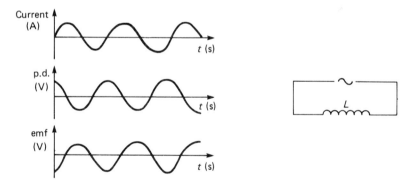

Figure 8.72

In this case the current *lags* behind the p.d. by $\pi/2$. When the current is zero, there is no magnetic flux and so the rate of increase of current is a maximum.

From the equation for *self-inductance* L the induced emf in the coil itself (called the back emf) is

$$\varepsilon = - L \frac{dI}{dt}$$

and this is therefore a *maximum* for *zero current*.

From *Lenz's law* the *back emf* in the inductor must be equal but opposite in direction to the applied p.d.

For maximum current $dI/dt = 0$ and therefore both back emf and applied p.d. are zero. So

$$v = -\varepsilon = L\frac{d}{dt}(i_0 \sin \omega t)$$

and

$$\boxed{v = \omega L i_0 \cos \omega t}$$

Reactance

This is the opposition of an inductor or capacitor to a.c. and is defined by

$$\boxed{X = \frac{v_0}{i_0}}$$

For capacitors,

$$X_C = \frac{v_0}{\omega C v_0} \quad \text{and so} \quad \boxed{X_C = \frac{1}{\omega C}}$$

For inductors,

$$X_L = \frac{\omega L i_0}{i_0} \quad \text{and so} \quad \boxed{X_L = \omega L}$$

The unit of reactance is the **ohm** (symbol Ω).

Impedance, *Z*

This is the opposition of *a circuit* to a.c.

For a *resistor–capacitor series* circuit,

$$Z = \frac{v}{i} = \sqrt{R^2 + X_C^2}$$

For a *resistor–inductor series* circuit,

$$Z = \frac{v}{i} = \sqrt{R^2 + X_L^2}$$

For a *resistor–capacitor–inductor series* circuit,

$$Z = \frac{v}{i} = \sqrt{R^2 + (X_L - X_C)^2}$$

The unit of impedance is also the **ohm** (symbol Ω).

Resonance in *L–C–R* series circuits

Now

$$X_L \propto f \quad \text{and} \quad X_C \propto 1/f$$

and R does not change with f, where f is the frequency of a.c. used. Using a common y-axis the graph in Fig. 8.73 can be obtained. f_0 is called the **resonant frequency**.

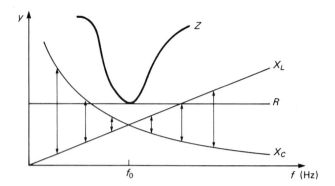

Figure 8.73

Vectorial representation of phase difference

The *phase difference* between p.d.'s (and indeed currents) discussed previously can be represented by vector diagrams. In Fig. 8.74 the phase of the current is taken as the reference line.

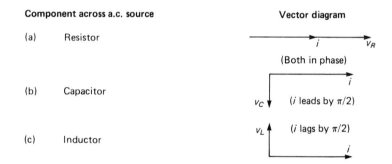

Figure 8.74

Since the phase of the current is the same as that of the p.d. across a resistor on a.c., Figs 8.74(b) and (c) can be used to represent the phase difference between the p.d. across a capacitor and the p.d. across an inductor as compared to that across the resistor for *C–R* and *L–R* series circuits *respectively* (Fig. 8.75).

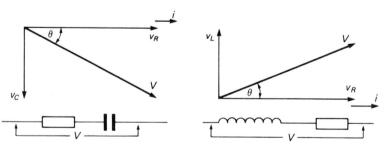

Figure 8.75

The phase difference between the total p.d. across the components in series V and the total current i is then θ. (The value of V can be obtained by Pythagoras' theorem, the formulae being given earlier.)

$$\tan \theta = -\frac{v_C}{v_R} = \frac{-iX_C}{iR} = \frac{-X_C}{R}$$

(negative sign indicates lag) and

$$\tan \theta = \frac{v_L}{v_R} = \frac{iX_L}{iR} = \frac{X_L}{R}$$

The situation for the *L–C–R* series circuit is represented in Fig. 8.76.

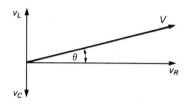

Figure 8.76

$$\tan \theta = \frac{v_L - v_C}{v_R} = \frac{X_L - X_C}{R}$$

Resonant frequency

At the *resonant frequency* f_0 the *impedance* of an *L–C–R* series circuit is a *minimum* and *equal* to the *pure resistance R* of the circuit.

This is because

$$X_L = X_C$$

and so

$$(X_L - X_C)^2 = 0$$

As a result of the minimum impedance, the current is a maximum (Fig. 8.77), and since the circuit is acting like a pure resistance the current and p.d. are *in phase*. (Hence the term **resonance**.)

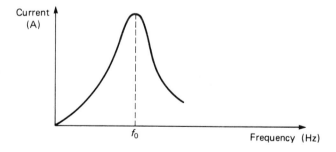

Figure 8.77

The *resonant current* has value

$$\boxed{i_r = \frac{v}{R}}$$

and since $X_L = X_C$

$$2\pi f_0 L = \frac{1}{2\pi f_0 C}$$

Therefore

$$f_0 = \frac{1}{2\pi\sqrt{LC}}$$

Tuning circuits

One of the consequences of resonance is the amplification of the p.d. across the capacitor or inductor, sometimes by 200 times. It is this amplification at or near the resonant frequency which is used in *tuning circuits*.

TV signals induce emf's of various frequencies in an antenna, which in turn induce currents of the same frequency in *L* by *mutual induction*. By *varying the capacitance* ('tuning in') the resonant frequency is also varied. See Fig. 8.78.

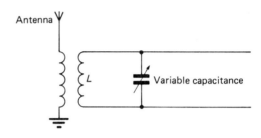

Figure 8.78

Example 8.15

Determine (i) the rms value and (ii) the peak value of a.c. which produces the same brightness for a bulb as when it is used on a 14 V d.c. supply.

(i)   From the definition of rms value, the rms value of the a.c. in this example will also be 14 V.

(ii)   If the peak value of the a.c. is v_0, then the rms value v is given by the expression:

$$v = v_0\sqrt{1/2}$$

Therefore

$$v_0 = v\sqrt{2} = 14\sqrt{2}$$

Thus,

$$v_0 = 19.8\ V$$

This example emphasises the fact that the term 'rms value' can be applied not only to current but to p.d. as well.

Example 8.16

Determine the maximum charge which passes to a 65 μF capacitor when 12 V rms is applied across it, and the rms current, if the frequency of this supply is 90 Hz.

Applying the equation $Q = CV$, then

$$Q_{max} = CV_{max}$$

The maximum value of the p.d. will be the peak p.d., v_0, and so

$$V_{max} = v_0 = v_{rms}\sqrt{2} = 12\sqrt{2} = 17\ V$$

Thus,

$$Q_{max} = 65 \times 10^{-6} \times 17 = \underline{1.11 \times 10^{-3}\,C}$$

The capacitor reactance

$$X_C = \frac{1}{2\pi fC} \quad \text{and} \quad X_C = \frac{v_{rms}}{i_{rms}}$$

Therefore

$$i_{rms} = \frac{v_{rms}}{X_C} = v_{rms}\,2\pi\,fC = 12 \times 2\pi \times 90 \times 65 \times 10^{-6}$$

Thus,

$$i_{rms} = \underline{0.44\,A}$$

Example 8.17

A 200 Ω resistor and a capacitor C are connected to form a series circuit with a 14 V rms a.c. supply of frequency 90 Hz. If the p.d. across each component is 9 V rms, calculate the value of C.

See Fig. 8.79.

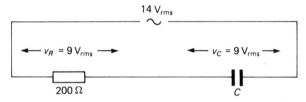

Figure 8.79

For the p.d. across the resistor v_R then

$$v_R = iR$$

Similarly, for the p.d. across the capacitor v_C then

$$v_C = iX_C$$

Thus,

$$X_C = v_C/i = v_C R/v_R$$

But also

$$X_C = 1/2\pi\,fC$$

Therefore

$$v_C R/v_R = 1/2\pi fC$$

Rearranging,

$$C = \frac{v_R}{2\pi f v_C R} = \frac{9}{2\pi \times 90 \times 9 \times 200} = \underline{8.8\,\mu F}$$

Example 8.18

The a.c. supply in the circuit of Fig. 8.79 is replaced with a 4 V d.c. battery and a switch. What is the initial rate of change of p.d. across the capacitor with respect to time, when the switch is closed?

The charge which can be delivered to the capacitor plates by the supply p.d., V, is given by

$$Q = CV$$

It can only do this initially by using the full current I. (Remember the current will die away as the capacitor is charged.) Therefore

$$It = CV$$

but

$$I = V/R$$

Thus,

$$\frac{V}{RC} = \frac{V}{t}$$

So

$$\frac{V}{t} = \frac{4}{200 \times 8.8 \times 10^{-6}} = \underline{2.3\,\mathrm{kV\,s^{-1}}}$$

Example 8.19

The series circuit in Fig. 8.80 has an a.c. supply of 14 V with a frequency of 90 Hz. The current in the circuit is out of phase with the p.d. by 55° and has a value of 0.4 A. (i) Determine the values of the resistance R and the inductance L. (ii) Determine the value of a capacitor C which, when placed in series with the other two components, gives rise to resonance in the circuit.

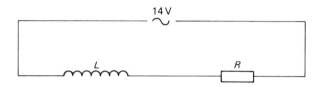

Figure 8.80

(i)   If the phase angle is ϕ then

$$\tan \phi = X_L/R$$

Thus,

$$R \tan \phi = X_L$$

Now

$$i = \frac{v}{Z} = \frac{v}{\sqrt{R^2 + X_L^2}} = \frac{v}{\sqrt{R^2 + R^2 \tan^2 \phi}}$$

Then

$$R = \frac{v}{i\,\sqrt{1 + \tan^2 \phi}} = \frac{14}{0.4\,\sqrt{1 + \tan^2 55°}}$$

Therefore

$$R = \underline{20\,\Omega}$$

Since

$$X_L = 2\pi f L = R \tan \phi$$

then

$$L = \frac{R \tan \phi}{2\pi f} \doteqdot \frac{20 \times \tan 55°}{2\pi \times 90} = \underline{0.05\,\mathrm{H}}$$

(ii)   For resonance to occur in the circuit then

$$X_L = X_C$$

and so

$$2\pi f L = 1/2\pi f C$$

Therefore

$$C = 1/4\pi^2 f^2 L$$

Thus,

$$C = 1/(4\pi^2 \times 90^2 \times 0.05) = \underline{63\ \mu F}$$

Deflection by parallel plates

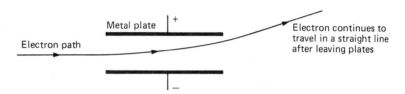

Figure 8.81

If two parallel metal plates have a p.d. applied across them (Fig. 8.81), then an **electric field** is set up between the plates. An electron travelling through the gap separating them will be deflected towards the higher potential. As its direction is changing so is its velocity. Therefore the electron is *accelerated* by the electric field.

Cathode ray oscilloscope (CRO)

The **oscilloscope** is a *thermionic* device, consisting of the following parts (see Figs 8.82 and 8.83).

(1)   An **electron gun** which produces a fine beam of fast electrons (cathode rays). These are emitted from a heated cathode C (see Fig. 8.82) into a cylinder called the grid G, which has a variable negative potential with respect to C. It therefore controls the number of electrons reaching the anode A_1 and controls the **trace brightness.**
    The anodes A_1 and A_2 *accelerate* the electrons in the evacuated glass tube as they both have a positive potential with respect to C. By adjusting the potential of A_1 the beam *can be focused.*

(2)   The **tube** itself is surrounded in mumetal which prevents the effects of external magnetic fields. It ends in a fluorescent screen, coated with zinc sulphide. On striking this screen the electrons cause *fluorescence*. A_2 is *earthed* to prevent the effect of earthed objects on the cathode rays.

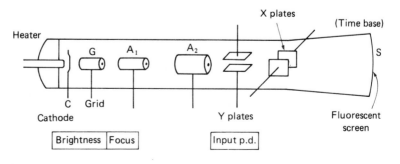

Figure 8.82

Typical frontal appearance

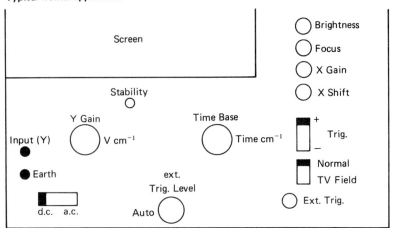

Figure 8.83

(3) **System of plates** The electrons emerge from A_2 with constant speed and pass between the horizontal Y plates and the vertical X plates. Due to p.d.'s across these plates the electrons are deflected vertically and horizontally respectively. By varying the potential on one of the Y and one of the X plates the spot formed on the screen can be shifted. Hence the *X and Y shift controls*.

(4) The **time base** If an alternating p.d. is applied across the Y plates, the beam is deflected continually up and down showing a vertical line trace on the screen, indicating the peak value of the alternating p.d. To see what happens to the alternating p.d. with time, the X plates are connected to a circuit called the **time base**. This provides a **saw-tooth p.d.** across the X plates.

The sawtooth p.d. of the time base has the appearance shown in Fig. 8.84.

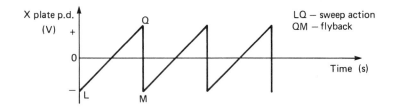

Figure 8.84

The *sweep action* 'pushes' the electron beam from side to side (i.e. from left to right), and because the p.d. changes dramatically at Q (see Fig. 8.84), the beam *flies back* to the left and re-starts the sweep. The combination of the effects of the X and Y plates gives rise to the trace observed, e.g. a sinusoidal one for alternating p.d.'s. Part of the alternating p.d. being observed is used to jolt or *trigger* the time base into sweep. The point at which this occurs can be varied by the *trig. level control*, although usually an automatic trigger is used.

Applications of the CRO

Voltmeter

Both a.c. and d.c. p.d.'s can be measured by applying them across the Y plates with zero time base. With d.c. the spot on the CRO screen is deflected vertically and, depending upon the setting of the Y gain control ($V\,cm^{-1}$), the p.d. can be determined by measuring the distance of displacement. A vertical line is obtained with a.c., whose length represents twice the *peak value* of the p.d. On application of the time base this vertical line expands laterally revealing how the p.d. varies with time. (School power packs set to d.c. *do not* give a deflection of the spot, but instead produce a vertical line. This is because pure, stable d.c. is not provided by them.)

Advantages over conventional voltmeters are as follows.

(1) Both a.c. and d.c. p.d.'s can be measured.
(2) Cathode rays act as a pointer of very small mass and almost instantaneous deflection.
(3) It has a very high resistance and impedance and so draws little current from the circuit it is connected to.

Clock

When the fine time base control (VARIABLE) is on CAL and the X gain is a *minimum*, the time base marked time per cm can be used as a clock. For example, on the 10 ms setting, the beam moves from left to right a distance of 1 cm in 10 ms. Now the frequency of mains is 50 Hz, i.e. there is one complete cycle every 0.02 s, and thus, on the 10 ms setting, one cycle of the trace would stretch across 2 cm on the screen.

Display unit

A waveform can be displayed by applying the observed p.d. across the Y plates and using the time base. The *phase difference* between two sinusoidally varying p.d.'s, one applied across the Y plates and the other across the X plates, can also be shown. If the p.d.'s have the same frequency the traces shown in Fig. 8.85 will be seen.

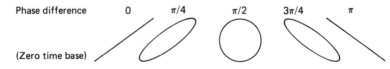

Phase difference   0 $\pi/4$ $\pi/2$ $3\pi/4$ π

(Zero time base)

Figure 8.85

These shapes are called **Lissajous figures**, although the term is usually associated with more complex patterns. The more complex patterns arise when two sinusoidally varying p.d.'s having *different frequencies* are applied to the X and Y plates.

9
The atom

Introduction

The work in this chapter divides itself roughly into two parts. The first considers the electron—its emission from a hot cathode, how it is influenced by electric and magnetic fields and the determination of its specific charge. This is followed by the determination of its actual charge and a discussion of black body radiation—an effect which together with the photoelectric effect and Planck's theory indicates how electrons are involved in the emission and absorption of electromagnetic radiation.

The second part, which starts off with radioactivity, deals with the nucleus. It discusses the properties and types of radiation which arise from the nucleus and how they are experimentally investigated. The basic structure of the atom and the stability of the nucleus follow. The chapter is concluded by a discussion of mass spectroscopy and electronic energy levels.

Revision targets

The student should be able to:

(1)   Describe the basic properties of cathode rays.
(2)   Explain qualitatively and quantitatively the effects of magnetic and electric fields upon an electron beam.
(3)   Define the term specific charge.
(4)   Describe methods for the determination of the specific charge and deduction of the actual charge of the electron.
(5)   Describe the basic properties of thermal radiation and explain Prèvost's theory of exchanges.
(6)   Define the term black body.
(7)   State Stefan's law and apply it to numerical problems.
(8)   State Wien's displacement law and apply it to numerical problems.
(9)   Describe the photoelectric effect and define work function.
(10)   Explain Planck's theory.
(11)   Apply Einstein's photoelectric equation to the solution of numerical problems involving the photoelectric effect.
(12)   Describe the basic properties and types of radiation.
(13)   Discuss the main types of radiation detector.
(14)   Explain, with reference to the Geiger–Marsden experiment, the basic construction of the atom.
(15)   Define proton and nucleon number, and the term isotope.
(16)   Say what the positron is.

(17)   Describe and give examples of the different types of radioactive decay.

(18)   Define the term binding energy.

(19)   Calculate and define the mass defect for a given nucleus.

(20)   Appreciate the relationship between stability of a nucleus and binding energy.

(21)   Discuss the process of fission, and define half-life.

(22)   Apply the decay law to the determination of half-life in numerical problems.

(23)   Make deductions about the stability of a nucleus from N–Z curve for stable nuclei.

(24)   Describe the determination of half-life for radon-220 gas.

(25)   Define relative atomic mass and calculate the relative abundances for naturally occurring isotopes.

(26)   Describe the Bainbridge mass spectrometer and perform calculations in simple problems involving it.

(27)   Discuss the concept of electronic energy levels and the experimental evidence for it.

Electron beams

When a metal is heated, some of the free electrons in it can travel to an extent when they have enough energy to escape the metal's surface. This is called **thermionic emission**, and the **work function** W of a metal is the energy just needed for an electron to escape the metal's surface. A useful unit of energy is the **electron-volt** (symbol eV). As the emitted electrons are often attracted to a plate at a positive potential with respect to the metal, the metal is called a **hot cathode**. If a.c. is to be used for heating, indirect heating is needed. (Why?)

Properties of cathode rays

A **cathode ray** is a beam of fast moving electrons and the *Maltese-Cross tube* (Fig. 9.1) can be used to demonstrate some of the cathode ray properties.

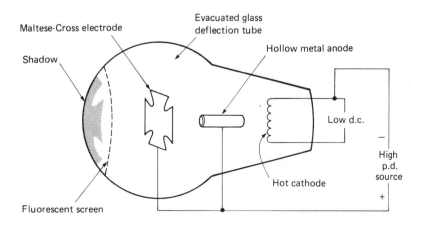

Maltese-Cross electrode

Shadow

Evacuated glass deflection tube

Hollow metal anode

Low d.c.

High p.d. source

Hot cathode

Figure 9.1   Fluorescent screen

In the tube, cathode rays pass through the hollow anode. If they miss the cross, they cause the screen to fluoresce, producing a *shadow*. Thus, cathode rays travel in *straight lines*.

Bringing up a N-seeking pole of a permanent magnet to the side of the tube causes the shadow to move upwards, but a dark *optical shadow* due to light from the hot cathode is unaffected. Thus, cathode rays are deflected by *magnetic* fields. The direction which the cathode rays take in the magnetic field is determined by applying Fleming's left hand rule. The cathode rays are also deflected by an *electric* field.

The paths of the cathode rays can be made visible by putting a small amount of hydrogen gas in the tube. This is *ionised* by the fast electrons, but recapture of freed electrons does take place with the emission of light.

As they are able to cause ionisation and *warm up* the end of the deflection tube, the cathode rays must *carry energy*. This energy is liberated by the production of X-rays and sometimes fluorescence when the rays strike a substance.

Work W is done on an electron in *accelerating* it from a hot cathode to an anode which has a p.d. of V. Thus, electrical potential energy is converted to kinetic energy of the electron. Hence

$$W = QV = \tfrac{1}{2}mv^2$$

and therefore

$$v = \sqrt{2QV/m}$$

For an electron, charge $Q = e$. Thus,

$$v = \sqrt{2eV/m}$$

Effect of a magnetic field

The force on an electron moving *perpendicularly* to a field of magnetic flux density B is

$$F = BeV$$

and its direction is given by Fleming's left hand rule. Both the direction of the motion and the field are always *perpendicular* to the force and the electron *speed* remains *constant*. Thus, the electron travels in a *circle* of radius r (Fig. 9.2).

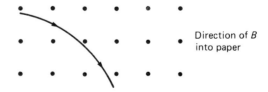

Direction of B into paper

Figure 9.2

(As the speed of the electron remains constant its energy does not change.) The force Bev is then the centripetal force

$$Bev = \frac{mv^2}{r}$$

and this acts towards the circle's centre.

Effect of an electric field

From the definition of electric field strength, the force on an electron travelling *perpendicularly* to the electric field is

$$F = Ee$$

where E is the electric field strength. Therefore acceleration

$$a = \frac{Ee}{m}$$

Both the force and the acceleration are directed towards the positive plate.

Vertical displacement:

$$y = \tfrac{1}{2}at^2$$

(since the vertical component of velocity v is zero).

Horizontal displacement:

$$x = vt$$

(since the horizontal component of acceleration a is zero).

$$\boxed{y = \tfrac{1}{2}\frac{ax^2}{v^2}}$$

Since both a and v are constant, the electron's path between the plates is a *parabola*.

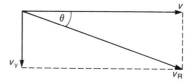

Figure 9.3

Due to the acceleration directed towards the positive plate, the electron acquires a velocity v_y towards the *positive plate* (Fig. 9.3). The resultant velocity v_R due to the constant velocity v and the velocity v_y is that which the electron has as it *leaves* the plates. It then travels in a *straight line* with velocity v_R.

For an electron travelling *perpendicularly* to an electric field, between two plates:

$$\frac{change \text{ in velocity}}{\text{towards positive plate}} = v_y = at$$

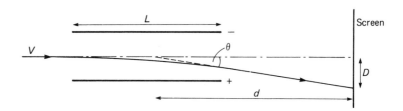

Figure 9.4

The time which it takes the electron to pass through the plates is L/v (Fig. 9.4), and therefore

$$v_y = \frac{aL}{v} = \frac{EeL}{mv}$$

If θ is the angle between the original path of the electron and the path it follows on leaving the plates then, with reference to the vector diagram in Fig. 9.3,

$$\tan \theta = \frac{v_y}{v}$$

so $$\tan \theta = \frac{EeL}{mv^2}$$

but also

$$\tan \theta = \frac{D}{d}$$

If before entering the plates a p.d. of V has been used to accelerate the electrons to the speed v, then the work done is

$$eV = \tfrac{1}{2}mv^2$$

and substituting,

$$\tan \theta = \frac{EL}{2V}$$

Finally, since $\tan \theta = D/d$,

$$\boxed{D = \frac{dEL}{2V}}$$

Example 9.1

An electron is accelerated by a p.d. of 400 V towards a pair of parallel metal plates 3 cm long separated by an air gap of 1 cm. The electron enters midway between the plates which are maintained at a p.d. of 60 V and on leaving them continues with a constant velocity until it strikes a vertical, fluorescent screen. If the screen is 18 cm from the centre of the air gap, find (i) the work done in initially accelerating the electron, (ii) the speed at which the electron enters the parallel metal plates, and (iii) the vertical displacement away from its original path after passing through the plates and striking the screen. (Electronic charge and mass are 1.6×10^{-19} C and 9.1×10^{-31} kg respectively.)

(i)   The work done in initially accelerating the electron is

$$W = eV = 1.6 \times 10^{-19} \times 400 = \underline{6.4 \times 10^{-17}\,\text{J}}$$

(ii)   On being initially accelerated the electron gains kinetic energy given by

$$\text{KE} = eV = mv^2/2$$

Thus,

$$v^2 = 2eV/m$$

Therefore

$$v = \sqrt{\frac{2 \times 1.6 \times 10^{-19} \times 400}{9.1 \times 10^{-31}}} = \underline{1.19 \times 10^7 \, \text{m s}^{-1}}$$

(iii)   The electric field strength of the field between the parallel metal plates is

$$E = \frac{V_{\text{p}}}{s} = \frac{60}{0.01} = 6 \, \text{kV m}^{-1}$$

where s is the plate separation. Thus, the vertical displacement away from the original path of the electron on striking the screen is

$$D = dEL/2V = (0.18 \times 6 \times 10^3 \times 0.03)/(2 \times 400) = 0.041 \, \text{m}$$

i.e. $D = \underline{4.1 \, \text{cm}}$

Crossed fields method for e/m

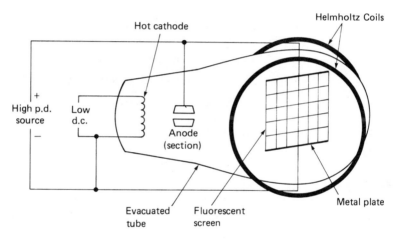

Figure 9.5

Electrons emitted from the hot cathode (Fig. 9.5) are accelerated by the anode down the evacuated glass tube. A *horizontal slit* in the anode forces the electrons into a flat beam which strikes the fluorescent screen. This screen is set *vertically* between two metal plates *at an angle* to the beam. An electric field is provided by these plates of separation d when a p.d. of V is applied across them. *Helmholtz Coils* either side of the tube provide a uniform magnetic field *perpendicular* to both the direction of electron motion and the electric field. If the fields are uniform, an electron is undeflected when the magnetic force equals the electric force, but their directions are opposite.

$$\boxed{Ee = Bev}$$

If V is also the accelerating p.d., the energy gained by the electrons is

$$\tfrac{1}{2}mv^2 = eV$$

and on substituting,

$$e/m = E^2/2B^2 V$$

From the definition of potential gradient, for a uniform field, the electric field strength of the field between the plates is

$$E = V/d$$

and so

$$\frac{e}{m} = \frac{V}{2B^2 d^2}$$

The experimental procedure is to obtain a deflection of the electron beam using the magnetic field *only*. The p.d. across the plates is then increased from zero until the deflection has been 'cancelled out'. The p.d. at which this occurs is noted from the eht unit meter. (This supplies the p.d. across the plates.) The value of the magnetic flux density is then calculated using

$$B = 0.72\,\mu_0\,NI/r$$

where N is the number of turns on one of the coils used, I is the current through the coil, and r is the coil radius. The permeability of free space is given by

$$\mu_0 = 4\pi \times 10^{-7}\,\mathrm{H\,m^{-1}}$$

Finally the plate separation d is measured and then the **specific charge** of the electron can be determined.

$$\boxed{e/m \text{ is about } 1.76 \times 10^{11}\,\mathrm{C\,kg^{-1}}}$$

The value of the electronic charge e can be accurately determined from electrolysis and hence the electron mass can be calculated by using the specific charge value.

Millikan's experiment

Faraday's constant F is the charge carried by a mole of any *monovalent* ions, and is about 96 500 C.

number of ions in a mole = *Avogadro's constant N* = 6×10^{23}

Therefore

each ion carries a charge (i.e. electronic charge e) $= \dfrac{96\,500}{6 \times 10^{23}} = \underline{1.6 \times 10^{-19}\,\mathrm{C}}$

A value for e can also be determined from *Millikan's* experiment. This uses charged oil drops which fall in air and are then raised or stopped by the application of an electric field.

Air fills a *gap* between two metal plates, one resting on a metal case, the other with a very small hole for a spray resting on some insulation. (See Fig. 9.6.) A *fine spray* of *oil droplets* is produced and a few descend through the hole into the air gap which is *strongly illuminated* from the side, for viewing with a *microscope*.

The time taken for a particular drop to descend through a certain distance on the *microscope scale* in the eyepiece, is measured. This scale has been *calibrated* using a millimetre scale. The *velocity* of that particular drop is then calculated.

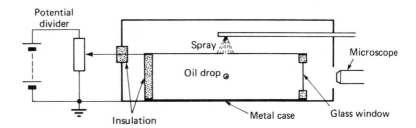

Figure 9.6

Assuming an oil drop is spherical, the resistance it experiences due to the air is

$$F = krv$$

where k is a *constant*, r is the radius of the drop, and v is the velocity of the drop.

As it is small, the drop travels with its *terminal speed* almost immediately, i.e. the resultant force on it is zero. Neglecting the upthrust on the drop due to the air, its weight must therefore equal F in the above equation. Its weight is given by

$$W = mg = V\rho g = \tfrac{4}{3}\pi r^3 \rho g$$

where ρ and V are the density of the oil and the volume of the drop respectively, and g is the acceleration due to gravity.

Therefore

$$\tfrac{4}{3}\pi r^3 \rho g = krv$$

Thus, r can be found if k and v are known. Friction or ions in the air will have *charged* the drop with a charge Q, say. To keep the drop *stationary*, a p.d. of V is applied (in the correct sense) across the plates. The force then on Q is

$$EQ = \tfrac{4}{3}\pi r^3 \rho g$$

where the electric *field strength* $E = V/d$ and d is the separation of the plates. Thus, Q can be found. It is always determined as a multiple of the *electronic charge e.*

Example 9.2

An oil drop carrying a charge Q is observed to fall in a Millikan's experiment with a steady speed of 5.7×10^{-4} m s^{-1}. If it is then held stationary when a p.d. of 3.2 kV is applied across the plates and assuming negligible upthrust due to the air, find (i) the radius of the oil drop and (ii) the value of Q. (Plate separation is 4 mm, density of oil is 830 kg m^{-3}, the constant of proportionality for the resistive force due to the air is $k = 4.2 \times 10^{-4}$ N s m^{-2}, and $g = 9.8$ m s^{-2}.)

(i)   When the oil drop moves with a steady speed (its terminal speed), the forces acting upon it are balanced. Therefore

$$W = \tfrac{4}{3}\pi r^3 \rho g = krv$$

where ρ is the density of the oil and r is the drop radius. Thus,

$$r = \sqrt{\frac{3kv}{4\pi \rho g}} = \sqrt{\frac{3 \times 4.2 \times 10^{-4} \times 5.7 \times 10^{-4}}{4\pi \times 830 \times 9.8}}$$

Therefore

$$r = \underline{2.66 \times 10^{-6}\,\text{m}}$$

(ii) When the drop is held stationary in the electric field between the plates,

$$EQ = \tfrac{4}{3}\pi r^3 \rho g$$

thus,

$$Q = \frac{4\pi r^3 \rho g}{3E}$$

Since $E = V/d$ then

$$Q = \frac{4\pi \times (2.66 \times 10^{-6})^3 \times 830 \times 9.8}{(3 \times 3200)/(4 \times 10^{-3})}$$

Thus,

$$Q = \underline{8 \times 10^{-19}\,\text{C}} \qquad (\text{i.e. } 5e)$$

Thermal radiation

This is a transfer of energy by electromagnetic waves, mainly in the form of infrared.

Wavelengths of infrared (IR)

Near infrared $0.75\,\mu$m to $10\,\mu$m.   Origin: molecular vibrations.

Far infrared $10\,\mu$m to $100\,\mu$m.   Origin: molecular rotations.

The main **properties** of infrared are as follows.

(1) It moves with the speed of light and travels in straight lines.
(2) It can travel in a vacuum.
(3) It can be reflected, refracted, polarised and undergoes interference.

Absorption and emission

Infrared, like visible light, is absorbed by dark, rough surfaces and reflected by light, smooth surfaces. This can be demonstrated by the use of *Leslie's cube* which is a hollow metal cube whose vertical sides are painted shiny black, dull black and white, the fourth being left silver coloured. The cube is filled with hot water and is then left for about ten minutes. A thermopile connected to a light beam galvanometer is next pointed at each face in turn and the deflections on the galvanometer represent how well the surface is emitting heat radiation.

Detection of infrared

(1) Ordinary sensitive mercury thermometer with blackened bulb.
(2) *Thermopile*—consists of a series of thermocouples connected in series. The *hot junctions* are blackened and exposed to the radiation while the *cold junctions* are shielded by a cap. A sensitive galvanometer is then placed in series with all the thermocouples (Fig. 9.7).

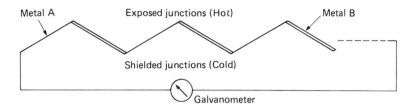

Figure 9.7

Prèvost's theory of exchanges

If a body's temperature is the same as that of its surroundings, the rate at which it emits radiation to them must be equal to the rate at which it absorbs radiation from its surroundings (provided convection and conduction are not also occurring).

Every body radiates energy at a rate controlled by its *thermodynamic temperature T*.

(1) If $T_{\text{body}} > T_{\text{surroundings}}$ there is a net loss of radiant energy from the body.
(2) If $T_{\text{body}} < T_{\text{surroundings}}$ there is a net gain by the body.
(3) If $T_{\text{body}} = T_{\text{surroundings}}$ *dynamic equilibrium* is reached and there is no net loss or gain from the body.

So a good absorber *is also* a good emitter. Thus, an ice-cold body radiates heat. A poor absorber is a poor emitter.

Black body

This is a *theoretical body* which absorbs all the radiation falling onto it. As it is the perfect absorber *it must also* be the perfect emitter. An approximation to a black body is a large, hollow copper sphere coated inside with platinum black, with a small hole.

Radiation entering the hole is trapped inside after only a few reflections. Similarly at *any temperature T*, the hole emits radiation of *all wavelengths*. This is called **black body radiation** and depends only upon *T*.

Stefan's law (Stefan–Boltzmann law)

The total energy radiated per unit area per second by a black body is proportional to the fourth power of the thermodynamic temperature of that body.

$$\varepsilon = \sigma T^4$$

The constant of proportionality is σ and is called **Stefan's constant**. Using Stefan's law, the rate of cooling of a black body is proportional to $(T^4 - T_0^4)$, where T_0 is the thermodynamic temperature of the surroundings. (Stefan's constant $\sigma = 5.7 \times 10^{-8} \, \text{W m}^{-2} \, \text{K}^{-4}$.)

Wien's displacement law

If the maximum energy emitted per unit area per second per unit wavelength ε_{max} arises at a wavelength λ_{am} for a black body at a thermodynamic temperature of T, then

$$\boxed{\lambda_{am} = constant/T}$$

where the *constant* is 0.003 mK. From Stefan's and Wien's laws, the maximum energy emitted per unit area per second per unit wavelength is

$$\boxed{\varepsilon_{am} \propto T^5}$$

See Fig. 9.8.

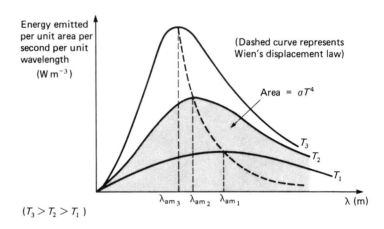

Figure 9.8

$(T_3 > T_2 > T_1)$

When metals are heated they glow red, then orange, yellow-white and possibly even blue as they get hotter. This is a good example of *Wien's law*. Stars are *approximately* black bodies but in astrophysics it is useful to calculate their 'black body temperatures' making the assumption that they are perfect black bodies (e.g. the Sun's black body temperature is 6000 K).

Example 9.3

Two cubes are made from the same material. They have similar surfaces but one cube has side L and the other $2L$. If the larger cube has a thermodynamic temperature three times that of the smaller cube, determine the ratio of the rate of emission of the larger cube to that of the smaller one. (Assume that heat absorption from the surroundings is negligible.)

From Stefan's law the rate of emission from surface area A is given by

$$R = \sigma T^4 A$$

Let R_1 and R_2 be the rates of emission of the large and small cubes respectively. Then

$$\frac{R_1}{R_2} = \frac{(3T)^4 \times 6 \times (2L)^2}{T^4 \times 6 \times L^2} = \frac{3^4 \times 6 \times 2^2}{6}$$

Therefore

$$R_1/R_2 = \underline{324}$$

That is, the larger cube emits radiation 324 times faster than the smaller cube.

Example 9.4

A metal cylinder is heated by passing an electric current through it. If it has a length $L = 0.87$ m and a diameter $d = 4$ mm, with a power output of 900 W, determine its working temperature assuming it to be a perfect black body. (Stefan's constant $\sigma = 5.7 \times 10^{-8}$ W m^{-2} K^{-4}.)

Let A be the surface area of the metal cylinder.

$$\text{power supplied to cylinder} = \text{power output of cylinder}$$

Therefore

$$P = A\sigma T^4 = \pi d L \sigma T^4$$

Thus,

$$T = \left(\frac{P}{\pi d L \sigma}\right)^{1/4}$$

Finally,

$$T = \left(\frac{900}{\pi \times 4 \times 10^{-3} \times 0.87 \times 5.7 \times 10^{-8}}\right)^{1/4}$$

Therefore

$$T = \underline{1096 \text{ K}}$$

(This value assumes negligible heat loss via processes other than radiation.)

Example 9.5

A cube which can be regarded as a perfectly black body has side L and radiates heat at 350 J per second to its surroundings which are at a temperature of 7 K. If the cube has a temperature of 280 K, determine its rate of emission when (i) the length of its side is doubled and (ii) the temperature of its surroundings is increased to 280 K.

(i) From Stefan's law

$$E = \sigma T^4 A$$

where E is the energy radiated by the cube per second and A is the surface area of the cube. Therefore

$$\frac{E_1}{E_2} = \frac{\sigma T^4 A_1}{\sigma T^4 A_2} = \frac{A_1}{A_2} = \frac{6 \times L^2}{6 \times (2L)^2} = \frac{1}{4}$$

Thus,

$$E_2 = 4E_1 = 4 \times 350 = \underline{1.4 \text{ kW}}$$

(ii) The rate of heat emitted by the cube is unaffected by the change in the temperature of the surroundings, and Stefan's law still applies.

Thus, the rate of heat radiated is once again 350 W assuming that no other changes, apart from the increase in temperature of the surroundings, have taken place.

If the surroundings have a temperature T_0, then the net heat radiated per second is given by

$$E = \sigma(T^4 - T_0^4)A = \sigma(280^4 - 280^4)A = 0.0 \text{ W}$$

Example 9.6

What would be the effect of increasing the temperature of the surroundings to 290 K in example 9.5, everything else being unaltered?

$$E_3 = \sigma(T^4 - T_0^4)A = \sigma(280^4 - 290^4)A = -9.3 \times 10^8 \sigma A$$

But

$$E_1 = 6.2 \times 10^9 \sigma A = 350\,\text{W}$$

therefore

$$E_3 = 0.15 E_1 = 53\,\text{W}$$

Thus, the body is absorbing heat at a net rate of 53 W.

Photoelectric effect

This is the ejection of electrons from metallic surfaces irradiated by *high frequency* electromagnetic radiation, e.g. ultraviolet (UV) or X-rays. Such metals include zinc, caesium and sodium. This effect can be demonstrated by exposing a clean zinc plate, resting on a *positively* charged electroscope, to ultraviolet radiation. Due to the positive charge, any ejected electrons are attracted back and the leaf remains undisturbed. If, however, the electroscope is *negatively* charged, the ejected electrons become repelled and the leaf collapses.

When glass is put between the electroscope and the source of UV, the collapse does not occur, i.e. UV does not penetrate glass.

Laws of photoelectricity

(1)   The number of ejected photoelectrons per second is proportional to the radiation *intensity*.

(2)   Photoelectrons are ejected with varying kinetic energy up to a *maximum* which increases with the frequency of radiation and does not depend upon the radiation intensity.

(3)   There is a minimum or *threshold frequency* of radiation for each metal below which photoelectrons are not ejected.

Planck's theory

This considers electromagnetic radiation not to be emitted continuously, but rather in 'packets' of energy which are called **quanta**. A source of electromagnetic radiation can emit several such 'packets' but not a fraction of a packet. Each quantum has energy

$$\boxed{E = hf}$$

where f is the frequency of radiation and h is called **Planck's constant** and has the value

$$h = 6.6 \times 10^{-34}\,\text{J s}$$

The higher the frequency, the greater is the energy. A quantum of electromagnetic radiation is known as a **photon**.

Einstein's photoelectric equation

The quantum theory can explain the laws of photoelectricity. When radiation of frequency f is shone at a metal, electrons are only ejected if hf is greater than the work function W of the metal. Any remaining energy supplied to the metal gives the electrons their kinetic energy outside the metal surface. Thus, the largest kinetic energy which an electron may possess is given by

$$\boxed{\tfrac{1}{2}mv^2_{max} = hf - W}$$

This is called **Einstein's photoelectric equation**.

Thus, the minimum or **threshold** frequency f_0 needed to eject an electron is given by

$$hf_0 = W$$

and so

$$\boxed{\tfrac{1}{2}mv^2_{max} = h(f - f_0)}$$

It can also be seen that the radiation frequency f increases as the maximum kinetic energy possible for a *photoelectron*.

Assume for the following that Planck's constant $h = 6.6 \times 10^{-34}\,\text{J s}$.

Example 9.7

If the speed of light is $3 \times 10^8\,\text{m s}^{-1}$, find the energy carried by a photon of (i) green light of wavelength 5.1×10^{-7} m, (ii) radio waves of wavelength 0.7 m (UHF), and (iii) gamma rays of wavelength 1.3×10^{-13} m.

From Planck's theory a photon carries energy

$$E = hf = hc/\lambda$$

(i) For green light,

$$E = \frac{6.6 \times 10^{-34} \times 3 \times 10^8}{5.1 \times 10^{-7}} = \underline{3.88 \times 10^{-19}\,\text{J}}$$

(ii) For radio waves,

$$E = \frac{6.6 \times 10^{-34} \times 3 \times 10^8}{0.7} = \underline{2.83 \times 10^{-25}\,\text{J}}$$

(iii) For gamma rays,

$$E = \frac{6.6 \times 10^{-34} \times 3 \times 10^8}{1.3 \times 10^{-13}} = \underline{1.52 \times 10^{-12}\,\text{J}}$$

Example 9.8

If the speed of light is $3 \times 10^8\,\text{m s}^{-1}$ and the mass of an electron is 9.1×10^{-31} kg, determine the following quantities for a metal used in a demonstration of photoelectricity, which has a threshold wavelength of 5.12×10^{-7} m: (i) the threshold frequency of the metal, (ii) its work function, (iii) the maximum photoelectron speed, and (iv) the maximum photoelectron energy when the metal is illuminated with radiation of wavelength 4.52×10^{-8} m.

(i) Applying the expression $v = f\lambda$, then

$$f_0 = c/\lambda_0$$

where the subscript '0' indicates a threshold value. Thus,

$$f_0 = (3 \times 10^8)/(5.12 \times 10^{-7}) = \underline{5.88 \times 10^{14}\,\text{Hz}}$$

(ii) The work function

$$W = hf_0 = 6.6 \times 10^{-34} \times 5.88 \times 10^{14}$$

Therefore
$$W = \underline{3.88 \times 10^{-19} \, \text{J}}$$

(iii)   From Einstein's photoelectric equation,

$$\tfrac{1}{2}mv_{max}^2 = hf - W$$

Thus,

$$v_{max} = \sqrt{\frac{2(hf - W)}{m}}$$

and

$$hf = \frac{hc}{\lambda} = \frac{1.98 \times 10^{-25}}{4.52 \times 10^{-8}}$$

Thus,

$$v_{max} = \underline{2.96 \times 10^6 \, \text{m s}^{-1}}$$

(iv) $E_{max} = \tfrac{1}{2}mv_{max}^2 = 0.5 \times 9.1 \times 10^{-31} \times (2.96 \times 10^6)^2$

Thus,

$$E_{max} = \underline{4 \times 10^{-18} \, \text{J}}$$

Radioactivity

The spontaneous disintegration of an atomic nucleus is called radioactivity. In the process, the following can be emitted: alpha rays; beta rays; gamma rays; X-rays and other emissions.

Characteristics of radioactivity

Some characteristics are as follows.

(1)   The rate of disintegration depends upon the number of *unchanged* nuclei present and upon the radioactive substance itself.
(2)   Decay is *random*.
(3)   The rate of disintegration does *not depend* upon external conditions, indicating the nucleus to be the active part of the atom.
(4)   The kinetic energy of the alpha and beta emissions can be used to provide heat energy for power stations.
(5)   Radiation will *expose* a photographic plate (or film), and *ionise* a gas.

The last point can be demonstrated by the discharge of a charged electroscope by a radioactive substance placed above it. The charge on the electroscope is neutralised as a result of the ionising effect of the radioactive substance in the air.

Types of radiation

Alpha

Alpha rays are stopped by a thick sheet of paper and have a range in air of a few cm. They *strongly* ionise gases and are deflected by electric and magnetic fields. This indicates they are fairly *heavy, positive* particles (direction of deflection). In fact, they constitute helium nuclei, i.e. they are a combination of two protons and two neutrons. They are all emitted from a substance with the same speed. Plutonium-239 is a source of alpha particles.

Beta

Beta rays can penetrate a few mm of aluminium foil and have a range in air of a few metres. They *weakly* ionise air and are deflected by electric and magnetic fields, such that they appear to be *light, negative* particles. Specific charge measurements indicate beta particles to be fast moving electrons. Strontium-90 is a source of beta particles.

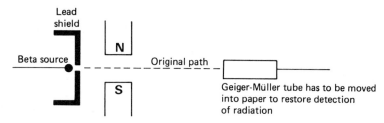

Figure 9.9

The deflection of beta particles in a magnetic field can be observed using the apparatus in Fig. 9.9. Lead shielding is used to prevent overload of the Geiger–Müller (G–M) tube. Initially the magnetic field is not applied and the beta particles go straight through. On application of the field, the G–M tube has to be *moved* to the side for *detection*.

Gamma

Gamma rays can penetrate a few cm of lead and *very weakly* ionise air. No deflection occurs in magnetic or electric fields. Gamma rays can be *diffracted* by crystals, travel in a vacuum with the speed of light and demonstrate the photoelectric effect. This indicates that they are *electromagnetic* waves of shorter wavelength than X-rays.

Radiation detection

Ionisation chamber

This is a gas-filled chamber with a metal rod at its centre. Radiation *ionises* the gas creating *ion-pairs*, i.e. electrons and positive ions. A *high p.d.* is applied across the chamber wall and the rod so that electrons are attracted to one and ions to the other, producing a current. This *ionising current* can be measured using an *electrometer/d.c. amplifier* and indicates the intensity of ionising radiation.

The set up shown in Fig. 9.10 can be used to measure the ionising current for alpha particles. (The electrometer/d.c. amplifier must first be *calibrated* by placing it across a potential divider in parallel with a voltmeter. When a 1 V reading on the voltmeter is obtained, the sensitivity control is adjusted until the moving coil meter scale reads a *full deflection*.)

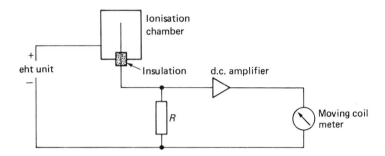

Figure 9.10

The p.d. across the ionisation chamber is increased until the moving coil meter reads a *maximum value*. Thus, the ionisation current $I = V/R$, where R is a very high resistance placed across the electrometer. The p.d. of V across R is obtained from the calibration of the moving coil meter.

Geiger–Müller tube (G–M tube)

This consists of an argon-filled cylindrical metal cathode with a central wire anode (Fig. 9.11).

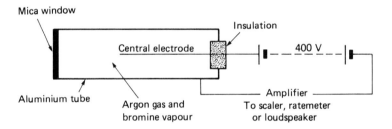

Figure 9.11

Beta and gamma radiation and very energetic alpha particles penetrate the mica window and *ion-pairs* are created in the gas. An electron from an ion-pair accelerates towards the anode (because of the p.d.) and on its way *collides* with argon atoms, releasing even more ion-pairs. The electrons from these in turn collide with more argon atoms, and so on, until an *avalanche* of electrons is produced at the anode.

This gives a *pulse of current*. In this way even a single ionisation can be detected by the G–M tube. After the avalanche the heavier positive ions accelerate towards the cathode but are prevented from causing an after-pulse by the use of *bromine vapour*. This process is called **quenching** and the bromine is called a **quenching agent**, although ethyl alcohol is sometimes used.

Quenching

As the positive ions move to the cathode they transfer their charge to the molecules of the vapour. These require less energy to be ionised than the argon atoms and the charged molecules move off to the cathode. The vapour molecules are broken up by electrons released on their arrival but because of their affinity for one another soon recombine.

Dead time

Since quenching is necessary, there is a dead time of about 200 μs between one *avalanche* and the next one which can occur. Penetrating radiation during that time will *not be registered*. So at most only 5000 ionising events per second can be counted by the G–M tube which, however, is sufficient for most purposes.

G–M tubes readily detect *beta particles* but need very thin mica windows to detect alpha particles. Gamma radiation is weakly ionising and so only about 1 % of it is detected by the G–M tube.

The pulses from the tube are *amplified* and recorded on a ratemeter or scaler. A *scaler* records the number of pulses in a certain time, whereas the *ratemeter* gives an average value for the number of pulses per second. A loudspeaker can be used to record the pulses as 'clicks'.

G–M tube characteristic curve

To find the *optimum working* p.d. of the G–M tube (about 400 V), a beta source is placed a short distance from it. Count rates are recorded as the p.d. for the tube is increased. Beyond the *Geiger threshold* on the plateau, the count rate is *almost constant*. All the ionising events are producing the same pulses and the avalanche occurs along the whole of the anode. For greater p.d.'s a damaging continuous discharge results instead of single pulses. See Figs 9.12 and 9.13.

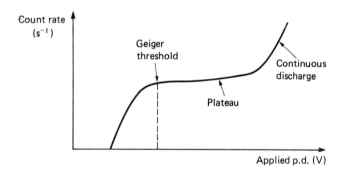

Figure 9.12

Figure 9.13

Background radiation

Radiation from *minerals* in the Earth, from contamination in the tube, and as a result of cosmic rays interacting with the Earth's atmosphere give rise to a count rate of about 25 per minute without the presence of a source. Therefore before using a source with the G–M tube it is necessary to measure the background count.

This count is then *subtracted* from subsequent counts using the source.

Expansion cloud chamber

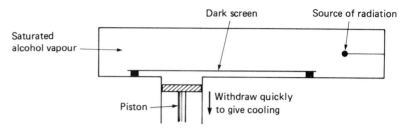

Figure 9.14

When the air within the chamber shown in Fig. 9.14 is ionised by the source and *cooled by rapid withdrawal* of a piston, alcohol vapour condenses onto the air ions. The white line of liquid drops, when illuminated against the screen, is called a *track*. The heavy alpha particles travel straight through, tearing electrons off air molecules. Beta particles suffer repulsion from electrons on atoms in the air, while gamma rays pass through occasionally interacting with atoms on the way.

Atomic structure

An idea of atomic structure can be deduced by targeting alpha particles at a sample of gold foil. This procedure is called the Geiger–Marsden experiment.

Geiger–Marsden experiment

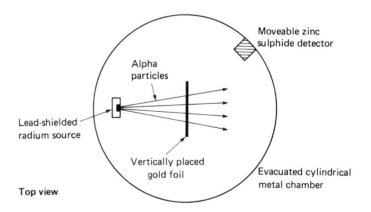

Figure 9.15

See Fig. 9.15. Alpha particles emitted by the radium pass *mainly undeflected* through the gold foil. They are detected by causing flashes of light on a *zinc sulphide* screen. Some alpha particles are deflected considerably, and about one in eight thousand is deflected back in the direction of the radium source.

It can be concluded from this that the atom consists of a *compact positive* nucleus which contains most of the *mass* of the atom—compact because only a few alpha particles are deflected, and containing most of the mass because the deflections are considerable. The nucleus is enveloped by a *distant* 'cloud' of electrons, the electrons being distant because they do not act as a shield against the alpha particles.

The theory of the experiment indicated a value for the nuclear radius to be of the order of 10^{-15} m as compared to 10^{-10} m for that of the whole atom.

This suggests that an atom is almost completely *empty space*. For a neutral atom the number of protons equals the number of electrons.

The neutron

Bombardment of beryllium with alpha particles produces radiation which can eject protons from paraffin wax and can also penetrate several cm of lead. This radiation cannot be deflected by magnetic or electric fields, and on the basis of conservation of mass and energy consists of a stream of neutral particles of mass almost equal to that of a proton. These are neutrons.

The nucleus

The nucleus is now considered to be made up of neutrons and protons. Any particle which resides in the nucleus is called a *nucleon*.

Wave mechanics

This considers electrons to have a wave-like nature and to exist in definite *energy levels* between which they can move by the emission or absorption of photons.

Proton number Z (atomic number)

This is the number of protons in an atomic nucleus. The number of electrons in a neutral atom equals Z. (It is the electrons which give an element its chemical properties.)

Nucleon number A (atomic mass)

The nucleon number specifies the number of nucleons in an atomic nucleus.

Element symbols

$^{4}_{Z}$Element abbreviation

For example:

$^{1}_{1}$H $\qquad$ $^{4}_{2}$He $\qquad$ $^{7}_{3}$Li

Isotopes

An atom of an element containing the same number of protons but a different number of *neutrons* as another atom of the same element is termed an **isotope**. For example:

$_1^1\text{H}$ (99.985%)   Hydrogen
$_1^2\text{H}$ (0.015%)   Deuterium
$_1^3\text{H}$ (radioactive)   Tritium

The value in brackets represents the percentage occurrence by mass. Tritium does not occur naturally. From isotope to isotope, the nucleon number A changes but the proton number Z does not. It is the variation of the nucleon number and the relative occurrence of each isotope which results in fractional values of the nucleon number. For example:

$$^{1.00797}_{\quad 1}\text{H}$$

Apart from the naturally occurring isotopes there are about 1200 artificial ones obtained in radioactive decays, although there are some naturally occurring radioactive isotopes. Isotopes of the same element have the same *chemical* properties since they still possess the *same number* of *electrons*.

Applications of radioactive isotopes

These are used in industry to check thickness, monitor levels, survey pollution, etc. Radioisotopes can also take part in chemical reactions and can therefore trace the progress of a particular chemical through a system. They are also used in the treatment of cancer and the sterilisation of bandages, etc. The dating of archeological artefacts can be done using carbon-14. This is created in the atmosphere by the interaction of cosmic rays with nitrogen. It can then be taken up by plants and trees, but when these die they are unable to take up any more. The duration of decay of the remaining carbon-14 can then be measured.

The positron

This is a particle possessing all the properties of an electron, except that it has a *positive* charge.

It is said to be the **antiparticle** of the electron and gives rise to a special type of radioactive decay called **beta positive** decay. An example of this is given on p. 262. Its production is equivalent to the loss of a proton.

Radioactive decay

The **decay** or **daughter product** is the result when an atom emits an alpha or beta particle. The decay continues until a *stable* decay product is obtained. A series of decays from one element is called a **decay chain** or **series.**

Examples of alpha and beta decay are shown on p. 262. When an alpha particle is emitted, the nucleon and proton numbers fall by four and two respectively. However, when a beta particle is emitted in ordinary or beta *negative* decay, the proton number increases by one while the nucleon number remains unchanged.

A nucleus can emit more than one particle at any one time and these may be a combination of the different types.

Gamma decay

After a nucleus has ejected an alpha or beta particle or both, it is left in an **excited state** and sometimes emits a gamma ray photon to stabilise itself. No change occurs in either the nucleon or proton numbers.

Nuclear energy

Einstein's energy–mass relationship $\Delta E = \Delta mc^2$ implies that any reaction giving rise to a considerable mass change is a potential source of energy. It also means that a unit of energy (e.g. the eV) can be thought of as a *unit of mass*. Mass in atomic Physics is often expressed in unified atomic mass units (symbol u, where 1 u is a twelfth of the mass of a *carbon*-12 atom).

For example $1\,eV = 1.6 \times 10^{-19}$ J can be considered as being equivalent to a mass change of 1.78×10^{-36} kg or 1.07×10^{-9} u.

Binding energy

For a particular atom, the sum of the masses of individual nucleons is *greater* than the mass of the nucleus. This difference in mass is called the **mass defect**. Einstein's energy–mass relationship shows that this loss in mass when a nucleus is formed from individual particles is equivalent to an energy gain. This energy is termed the **binding energy** of the nucleus.

For example consider an atom of oxygen-16. In terms of the unified atomic mass unit, the constituent particles of the atom give rise to the following masses:

8 protons + 8 electrons give a mass of $8 \times 1.007\,825$ u

8 neutrons give a mass of $8 \times 1.008\,665$ u

Thus, the total mass of the constituent particles of oxygen-16 is the sum of the above two products, i.e. 16.131 92 u. However, the mass of the atom as determined experimentally is 15.994 92 u, giving a mass defect of 0.137 u and hence a binding energy of 128 MeV.

Binding energy per nucleon curve

In many cases the binding energy per nucleon is about 8 MeV. Nuclides in the centre of the plot (Fig. 9.16) have *greater stability* since their binding energies are greatest.

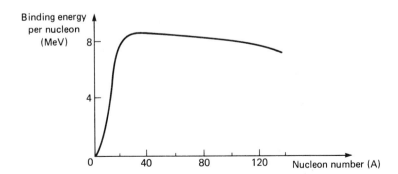

Figure 9.16

Binding energy is equal to the energy needed to split up the nucleus into its constituent particles. By *fusing* nuclei together or splitting them up to obtain nuclides located in the central region of the graph, greater binding energies are achieved. Sources of energy can thus be realised.

Fission

When a neutron is used as a 'bullet' particle targeted on uranium, the uranium nucleus can be split into *two large* nuclei. This process is called **fission**, and creates two fission fragments of nearly *equal mass*. Hence a considerable mass decrease occurs. In addition further neutrons are emitted. Several other heavy nuclides apart from uranium can be used in the same process.

An example of an interaction is

$$^{235}_{92}U + ^{1}_{0}n \rightarrow ^{141}_{56}Ba + 3^{1}_{0}n$$

In this interaction the *mass decrease* is about 10^7 greater than the equivalent energy change in a chemical reaction per atom. Most of the energy which results from the mass decrease takes the form of the kinetic energy of the *fission fragments*, and some the form of radiation. At least one of the fragments is radioactive.

A chain reaction can occur with the neutrons which are ejected in the fission process but, as many escape the surface of the target, a *minimum size* of target is needed for this to happen. A neutron source and two pieces of target material, which when brought together quickly just give this minimum size, form the basis of the atomic bomb.

The chain reaction can be applied more usefully in a nuclear reactor. The number of neutrons causing the reaction can be altered by the insertion of boron steel rods which absorb them. Uranium also contains the isotope *uranium-238* and graphite is used to slow neutrons down, preventing them being captured by it. High pressure gas is needed to transfer the heat generated by the reaction to water. This turns to steam and will drive the steam turbines in a power station. Radioisotopes are made by neutron bombardment of stable nuclides in nuclear reactors.

Decay law

This law is a statistical one which states that the decay of a radioactive substance occurs exponentially with time.

$$\boxed{N = N_0 e^{-\lambda t}}$$

where N is the number of nuclei present after time t, N_0 is the number of nuclei present when time $t = 0$ s, and λ is a constant with unit s^{-1}.

The law is based on the assumption that the rate of decay of a given substance at any time is directly proportional to the number of nuclei N present at the time considered.

$$\frac{-dN}{dt} = \lambda N$$

The constant of proportionality λ is called the **decay constant** and it is the fraction of nuclei present that decays per unit time.

Half-life $t_{1/2}$

This is the time taken for the number of active nuclei in a radioactive substance at a given time to be *halved*. The range of half-lives extends from microseconds to millions of years. For example:

Radioisotope	Half-life
$^{10}_{4}\mathrm{Be}$	25×10^6 years
$^{212}_{84}\mathrm{Po}$	3.04×10^{-7} s
$^{260}_{104}\mathrm{Ku}$	0.3 s

By plotting the percentage of active nuclei against time, for a particular radioactive substance, an *exponential decay* curve is obtained (Fig. 9.17).

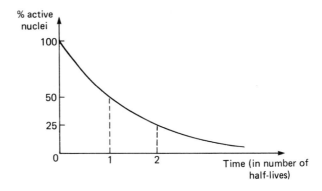

Figure 9.17

Since

$$N = N_0 e^{-\lambda t}$$

then

$$N_0/N = e^{\lambda t}$$

So for $t = t_{1/2}$ then $N = N_0/2$ and therefore

$$\boxed{2 = e^{\lambda t_{1/2}}}$$

so

$$\ln (2) = \lambda t_{1/2} \quad \text{or} \quad t_{1/2} = 0.693/\lambda$$

Example 9.9

The radioactive isotope neptunium-237 has a half-life of 2.14×10^6 years. Determine the time taken for its activity to be reduced by 67%.

From the expression $N = N_0 e^{-\lambda t}$, then for half life $t_{1/2}$,

$$\lambda = \frac{0.693}{t_{1/2}}$$

But

$$t_{1/2} = 2.14 \times 10^6 \text{ years} = 6.75 \times 10^{13} \text{ s}$$

Therefore, for neptunium-237,

$$\lambda = \frac{0.693}{6.75 \times 10^{13}}$$

So $\lambda = 1.03 \times 10^{-14}\,\text{s}^{-1}$

Now for the activity of the isotope to be reduced by 67%, the remaining activity must be 33% of its original value. Therefore

$$N/N_0 = e^{-\lambda t} = 0.33$$

Thus,

$$e^{\lambda t} = 1/0.33$$

and

$$t = \frac{\ln(1/0.33)}{\lambda} = \frac{1.1087}{1.03 \times 10^{-14}} = 1.076 \times 10^{14}\,\text{s}$$

So $t = 3.41 \times 10^6$ years

Example 9.10

The radioactive isotope manganese-54 has a half-life of 303 days. Determine the count rate measured from 0.004 kg of manganese. (Assume Avogadro's constant is $6 \times 10^{23}\,\text{mol}^{-1}$.)

The number of moles present in 0.004 kg of manganese is given by

$$\text{number of moles} = \frac{0.004}{0.054}$$

Therefore number of active atoms in 0.004 kg of manganese-54 is

$$N = \frac{6 \times 10^{23} \times 0.004}{0.054} = 4.44 \times 10^{22}\,\text{atoms}$$

Applying the expression $-dN/dt = \lambda N$, then

$$\frac{-dN}{dt} = \frac{0.693N}{t_{1/2}} = \frac{0.693 \times 4.44 \times 10^{22}}{303 \times 24 \times 3600}$$

Therefore the count rate is

$$\frac{-dN}{dt} = 1.17 \times 10^{15} \text{ counts per second}$$

The minus sign in the above answer indicates that the count rate has value which is falling with time.

N–Ż curve for stable nuclei

Atomic *stability* seems to depend upon the proton number Z and the number of neutrons N (Fig. 9.18).

The nucleus of an atom of a particular isotope is called a **nuclide**. Stable nuclides such as those of neon, calcium, tin and mercury lie on the stability curve. For lighter stable nuclides $N = Z$ *in general* and so the ratio $N/Z = 1$. For heavier stable nuclides more neutrons than protons are needed, and both N and Z are even for many nuclides which suggests the *stabilising* effect of alpha decay. (An alpha particle consists of two neutrons and two protons.)

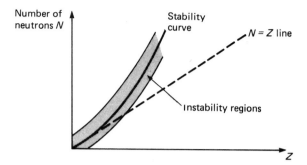

Figure 9.18

As unstable nuclides in the regions of instability around the curve decay they get progressively nearer the curve until they become stable. Those to the right of the stability curve possess an excess of protons or a lack of neutrons and thus decay by beta (+) decay or, for heavy nuclides, via alpha emission. Those unstable nuclides to the left of the curve possess an excess of neutrons or a lack of protons and will decay by beta emission.

Experimental determination of half-life

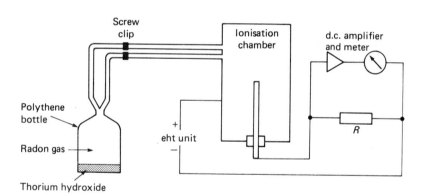

Figure 9.19

When the polythene bottle in Fig. 9.19 is squeezed, some radon-220 gas (sometimes called thoron) enters the ionisation chamber. The chamber is mounted on a d.c. amplifier and as the gas begins to decay the ionisation current is monitored. It is proportional to the quantity of gas remaining and the half-life may be found from a graph of meter reading versus time. Part of the decay which occurs is

$$^{228}_{90}\text{Th} \xrightarrow{\alpha} \,^{224}_{88}\text{Ra} \xrightarrow{\alpha} \,^{220}_{86}\text{Rn} \xrightarrow{\alpha} \,^{216}_{84}\text{Po}$$

The half-life of the gas is determined to be about 52 s.

Summary table of radiation properties

Property	α-radiation	β-radiation	γ-radiation
Symbol	$^4_2\alpha$	$^{\ 0}_{-1}\beta$	γ
Charge	$+2$	-1	Uncharged
Mass (amu)	4.0328	0.0006	—
Deflection in electric and magnetic fields	Slight and roughly equal deflection	Great, but to varying degree	No deflection
Speed of emission	About $2\times10^7\,\mathrm{m\,s}^{-1}$	Up to near $3\times10^8\,\mathrm{m\,s}^{-1}$	$3\times10^8\,\mathrm{m\,s}^{-1}$
Energy	Emitted from a source with nearly the same energy (4 MeV to 10 MeV)	Emitted from a source with different energies (up to 3 MeV)	Emitted with a few particular energies
Ionising ability	Strong	Weak	Very weak
Range in air	Few cm	Few metres	Many metres
Absorbed by	Thick paper	Few mm of aluminium foil	Few cm of lead
Electromagnetic wave	No	No	Yes

Summary table of particle properties

Particle	Symbol	Mass (amu)	Charge	Equation symbol
Proton	p	1.007 593	$+1$	$^1_1\mathrm{p}$
Neutron	n	1.008 982	0	$^1_0\mathrm{n}$
Electron	e^-	0.000 549	-1	$^{\ 0}_{-1}\mathrm{e}$
Positron	e^+	0.000 549	$+1$	$^{\ 0}_{+1}\mathrm{e}$

Examples of decay

Alpha
$$^{230}_{90}\mathrm{Th} \rightarrow\ ^{226}_{88}\mathrm{Ra} + ^4_2\alpha$$

Beta($-$)
$$^{234}_{90}\mathrm{Th} \rightarrow\ ^{234}_{91}\mathrm{Pa} + ^{\ 0}_{-1}\beta$$ A neutron is split into an electron (emitted) and a proton

Beta($+$)
$$^{30}_{15}\mathrm{P} \rightarrow\ ^{30}_{14}\mathrm{Si} + ^{\ 0}_{+1}\beta$$

Fusion

In this process energy is released when two *light nuclei* are brought together to give a heavier nucleus. It is the basis of the hydrogen bomb and the main source of energy in the Sun and other stars. An example of a reaction is

$$^2_1\mathrm{H} + ^2_1\mathrm{H} \rightarrow\ ^3_1\mathrm{H} + ^1_1\mathrm{H}$$

For each fusion the energy produced is 4 MeV. The reaction is *thermonuclear* as high temperatures are required to overcome the electrostatic repulsion of the two interacting nuclei: about a thousand million kelvin.

In the hydrogen bomb these high temperatures are achieved by the detonation of a small atomic bomb. Confinement of the high temperature gas involved, called **plasma**, in fusion reactions makes the development of a fusion reactor difficult. High currents passed through the deuterium provide the plasma while strong magnetic fields prevent it from touching the containing vessel. At present energy input for such a reactor is much greater than the energy output.

Radiation dangers

Exposure to external sources of radiation or the intake of radioactive substances by *ingestion* or *inhalation* is potentially dangerous. It can result in radiation burns followed by slow healing, blisters or sores, and in the long term cancers or cataract. Blood forming organs and the eyes are the most easily affected areas.

Relative atomic mass

This is given by

$$A_r = \frac{\text{mass of one atom}}{1/12 \text{ mass of carbon-12 atom}}$$

Modern *mass spectrometry* (the investigation of atomic masses) deals with many carbon-containing compounds and so carbon is a convenient *standard*. Also it has only two stable isotopes which almost always occur naturally in the same proportions.

Calculation of relative abundance

This will be treated in the form of an example.

Chlorine has a proton number of 17 and a relative atomic mass of 35.5. If there are two naturally occurring isotopes, chlorine-35 and chlorine-37, find the relative abundance of the isotopes.

Let the percentage abundance of $^{35}_{17}Cl$ and $^{37}_{17}Cl$ be a and b respectively. Then

$$a + b = 100\% \tag{1}$$

Now

$$\left(\frac{a}{100} \times 35\right) + \left(\frac{b}{100} \times 37\right) = 35.5$$

and substituting from equation (1) gives

$$\left(\frac{a}{100} \times 35\right) + \left(\frac{(100 - a)}{100} \times 37\right) = 35.5$$

Therefore

$$35a + 3700 - 37a = 3550$$

So $a = \underline{75\%}$ and, from equation (1), $b = \underline{25\%}$.

Mass separation and velocity selection

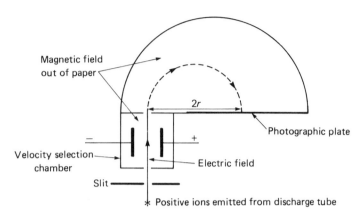

Figure 9.20

The device shown in Fig. 9.20 is called the **Bainbridge mass spectrometer**. It enables the mass of an individual atom to be obtained. Positive ions from the gas to be studied, possessing a range of speeds and charge to mass ratios, are channelled from a discharge tube into a beam by two slits.

They then enter crossed electric and magnetic fields, i.e. the fields are acting at right angles to each other.

If the magnetic and electric forces on an ion of charge Q are *equal*, then

$$BQv = QE$$

and so

$$\boxed{v = \frac{E}{B}}$$

where v is the speed of the ion. Thus, only undeflected ions have the same speed v. Only these therefore pass through the third slit, whereas deflected ions are unable to do so. This region of the spectrometer is called the **velocity selection chamber**.

After the third slit, only the magnetic field operates and the ions begin to travel in a circular path. Applying the idea of centripetal force, then

$$\boxed{BQv = \frac{Mv^2}{r}}$$

where r is the radius of the circular path and M is the mass of the ion under investigation. Therefore

$$r = \frac{ME}{QB^2}$$

since $v = E/B$.

Assuming the charge Q is the same for all the ions, and if E and B are constant and uniform,

$$r \propto M$$

In this way the mass of an *individual* atom can be found, whereas chemical experiments only give average values. By using an electrometer as a detector instead of the photographic plate, relative abundances can be determined from relative amounts of charge arriving at different points.

Example 9.11

The isotope magnesium-24 is ionised to form Mg^{2+} ions. These are then passed onto the velocity selection chamber of a Bainbridge-type mass spectrometer, where there is a uniform magnetic field of flux density 5.2×10^{-1} T. The plates in this part of the spectrometer are maintained at a constant p.d. of 400 V and have a separation of 4 mm. Determine (i) the electric field strength, (ii) the speed of the ions, and (iii) the kinetic energy of the ions, assuming that they emerge undeflected in the second chamber of the spectrometer. (Assume Avogadro's constant is 6×10^{23} mol^{-1}.)

(i)   Assuming no edge effects, i.e. considering a uniform electric field, then the electric field strength is given by

$$E = \frac{V}{d} = \frac{400}{4 \times 10^{-3}} = \underline{1 \times 10^5 \text{ V m}^{-1}}$$

(ii)   If the beam of ions is undeflected in the velocity selection chamber, then

$$v = \frac{E}{B} = \frac{1 \times 10^5}{5.2 \times 10^{-1}} = \underline{1.92 \times 10^5 \text{ m s}^{-1}}$$

(iii)   The kinetic energy of the ions is given by

$$KE = \tfrac{1}{2}mv^2$$

where m is the mass of each ion. But

$$m = M/N_A = 24/(6 \times 10^{23}) = 4 \times 10^{-23} \text{ g} = 4 \times 10^{-26} \text{ kg}$$

Therefore

$$KE = \frac{4 \times 10^{-26} \times (1.92 \times 10^5)^2}{2}$$

Thus,

$$KE = \underline{7.4 \times 10^{-16} \text{ J}}$$

Alternatively this can be expressed in terms of electron-volts. One electron-volt is the work done in accelerating unit charge through a p.d. of 1 V, i.e.

$$W = eV = 1.6 \times 10^{-19} \times 1 = 1.6 \times 10^{-19} \text{ J}$$

Therefore the kinetic energy of the ions is given in electron-volts from

$$KE = (7.4 \times 10^{-16})/(1.6 \times 10^{-19}) = \underline{4625 \text{ eV}}$$

Example 9.12

Determine (i) the specific charge and (ii) the path radius of the magnesium ions in example 9.11 when they enter the second chamber of the spectrometer, assuming the magnetic field there to be of the same flux density as in the first chamber.

(i)   The specific charge of the ions is given by

$$\frac{Q}{m} = \frac{2 \times 1.6 \times 10^{-19}}{4 \times 10^{-26}} = \underline{8 \times 10^6 \text{ C kg}^{-1}}$$

(ii)   Now,

$$r = \frac{mE}{QB^2} = \frac{1 \times 10^5}{8 \times 10^6 \times (5.2 \times 10^{-1})^2}$$

Thus,

$$r = \underline{4.6 \times 10^{-2}\,m}$$

Extra-nuclear atomic structure

Energy levels of an atom are the energies of electrons allowed by wave mechanics. They can be represented as shown in Fig. 9.21.

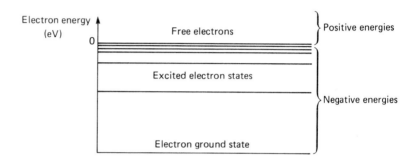

Figure 9.21

The atoms of different elements have different 'finger-print' sets of *energy levels*. Electrons are only allowed at these levels and *nowhere* in between. They move from one level to another by the loss or gain of the permitted amounts of energy called **quanta**. This occurs in the form of electromagnetic radiation—**photons**.

Conventionally an electron at rest just outside the atom is said to have **zero energy**. (See near the top of Fig. 9.21.) Any extra energy is said to be positive and takes the form of *kinetic energy*. If the electron is inside the atom it is said to have **negative energy**. The more negative, the lower is its energy. It can be thought of in terms of the electron being closer to the nucleus and requiring more energy to escape it than one further away.

The **ground state** is the lowest energy level in the atom and therefore the most stable since the electron can no longer lose energy. Electrons like to exist in the *lowest energy* state possible. If an electron is at a level from which it is able to 'fall' to one lower down, it is said to be in an **excited state**.

The **ionisation potential** (symbol I) is the energy required to remove the outermost electron from an atom to infinity. In the case of the energy-level diagram it is the energy difference between zero energy and the energy level of the outermost electron.

For example hydrogen has only one electron. It sits on the most stable energy level available, which is the ground state at $-13.6\,eV$. So the ionisation potential is $13.6\,eV$. Other ionisation potentials apply to the second and third, and so on, outermost electrons.

Shells are groups of electrons whose energies are almost the same.

Franck–Hertz experiment

This experiment provides practical confirmation for the idea of energy levels. In it electrons collide with atoms of a low pressure gas, when they have been accelerated by an applied p.d. When this p.d. is *equal* to the ionisation potential, the bombarding electrons supply enough energy for electrons to be knocked out of gas atoms by *inelastic collisions*.

When the accelerating p.d. is such as to cause the bombarding electrons to raise electrons in the gas atoms to a higher energy level, the p.d. is called the **excitation potential**. There are many of these for an atom and are particular to it, i.e. atoms can be identified by them. For elastic collisions only a small amount of bombarding electron energy is lost.

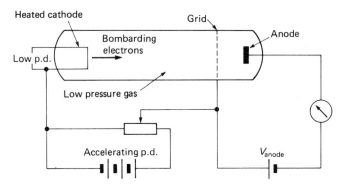

Figure 9.22

In the apparatus shown in Fig. 9.22, a *variable p.d.* accelerates electrons from the cathode to a wire grid through the low pressure gas. If the grid potential is just greater than the small negative potential that the anode has with respect to the grid, electrons reach the anode. A small current is then recorded by the galvanometer. This current increases as the accelerating p.d. increases and the bombarding electrons only suffer elastic collisions (Fig. 9.23).

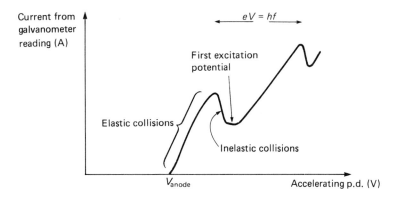

Figure 9.23

However, when the p.d. reaches a value called the **first excitation potential**, bombarding electrons suffer inelastic collisions. These cause electrons in the gas atoms to 'jump up' into their next possible energy states. As a result of the collisions, the incident electrons have lost most of their kinetic energy and are pushed back by the anode potential to the grid. A *drop in current* is then observed.

Further increases in the accelerating p.d. allow the electrons again to overcome the anode potential and current rises again. It is possible to continue the measurements further to determine other excitation potentials, and investigate ionisation potentials.

Ionisation potentials are low for elements with weakly bound outer electrons, e.g. the alkali and the alkali earth metals. They are higher for elements with almost complete or fully complete outer shells, e.g. the inert gases.

Spectra

Another effect which supports the concept of energy levels is the production of **spectra**. Optical spectra have wavelengths in the visible region of the electromagnetic spectrum and consist of two types—**emission** and **absorption** spectra. (See Chapter 10.)

10
Physical optics

Introduction

This chapter mainly covers optical instruments and devices, along with the theory associated with them. It begins with a consideration of the different types of spectra—also produced by other kinds of radiation besides light. Following this are sections on the spectrometer (which can be used to investigate spectra), and its applications. The topic of interference, which has been previously introduced, is then treated in more detail with particular reference to thin film interference.

The concepts of diffraction and interference come together in a discussion of the diffraction grating and its usefulness in wavelength determination. Light is part of a 'family' of radiations called the electromagnetic spectrum. The general and particular properties of these radiations are reviewed and this precedes a section dealing with the polarisation of light. Polarisation is a characterisitic property of all the electromagnetic radiations. The final part of the chapter concerns the refracting astronomical telescope and its features.

Revision targets

The student should be able to:

(1)   Discuss the production of, and difference between, emission and absorption spectra.
(2)   Relate the production of spectra to the extra-nuclear structure of an atom.
(3)   Describe the construction of the spectrometer and explain the adjustments required to prepare it for use.
(4)   Describe the way in which the spectrometer can be used to determine the angle of a prism, minimum deviation and the refractive index of the prism material.
(5)   Explain what is meant by the term interference.
(6)   Perform simple calculations involving Young's slits and describe how the slits can be used to determine the wavelength of monochromatic light.
(7)   Describe the interference of light in an air wedge and solve numerical problems concerning it.
(8)   Discuss the formation of coloured bands in thin films.
(9)   Qualitatively describe the formation and application of Newton's rings.
(10)   Explain what is meant by diffraction of light and distinguish it from interference.

(11)   Explain the features of Fraunhofer intensity patterns for single, double and multiple slit diffraction.

(12)   Apply the equation $n = d \sin \theta$ to numerical problems involving the diffraction grating.

(13)   Explain how a spectrometer-mounted diffraction grating can be used to determine the wavelength of monochromatic light.

(14)   Describe the main characteristics of the electromagnetic spectrum and the major properties of its main regions.

(15)   Explain what is meant by the polarisation of light and optical activity.

(16)   Describe the construction of the refracting astronomical telescope.

(17)   Define angular magnification and perform simple calculations involving the angular magnification of the astronomical telescope.

(18)   Define the term eye ring and explain what is meant by resolving power with reference to Rayleigh's criterion.

Emission spectra

Continuous spectra

Continuous spectra are produced by hot solids, liquids and gases at high pressures. (The radiation from a black body is said to form a *continuous* spectrum.) It is because the atoms are very close that all wavelengths are emitted and so the spectrum is *not particular* to the source. For example continuous spectra are produced by the Sun and a tungsten filament.

Most of the visible radiation emitted by the Sun is in the yellow-green region, to which the human eye is most sensitive—the yellow spot.

Line spectra

Line spectra are separate bright lines on a dark background, the lines having definite wavelengths. They are sometimes called **atomic spectra** and are produced by luminous gases and vapours at low pressures, in arcs and discharge tubes. They can also be seen in bunsen flames.

The line spectrum of an element completely *identifies it*. Emissions from atoms due to electron 'jumps' between energy levels provide the lines. Elements in distant stars can be identified by line spectra. The **Doppler effect** shows up in these stellar spectra. If the star is moving away, the lines are shifted towards the red end of the spectrum. This is the **red shift**. If the star is moving nearer, there is a **blue shift**. In this way the velocity of a star or gaseous cloud can be investigated. An example of a line spectrum is that of sodium where the characteristic yellow sodium D-lines occur (Fig. 10.1).

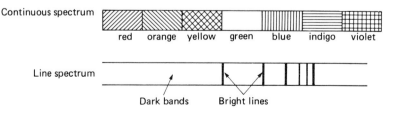

Figure 10.1

Absorption spectra

These are produced when radiation emitted is absorbed by material *between* the source and the observer. Line and continuous spectra are again obtained.

A line absorption spectrum consists of a continuous spectrum with dark lines across it. Atoms of the intervening material absorb certain wavelengths and re-emit them, but in *all directions*. So a weak emission spectrum on top of a continuous one gives the effect of *dark lines*.

Dark lines crossing the Sun's spectrum, which is continuous, are caused by layers of gas enveloping it. They are called *Fraunhofer lines*. Helium was first discovered not on Earth but in the Sun from the Fraunhofer lines in 1868.

Optical line spectra and energy levels

Each line in an optical line spectrum represents the movement from a higher excited energy state to a lower one, and a definite frequency of radiation emitted as a result (Fig. 10.2).

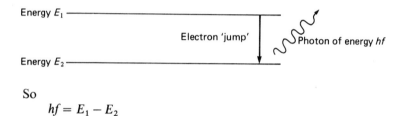

Figure 10.2

So
$$hf = E_1 - E_2$$
where h is Planck's constant and f is the frequency of radiation.

Hydrogen line spectrum

For hydrogen (Fig. 10.3), the optical line spectrum consists of a series of lines called the Balmer series which represents the transition of electrons down to the first excited energy state.

If the hydrogen line spectrum is extended into non-optical regions, similar series of lines are formed—the Lyman series (UV) and the

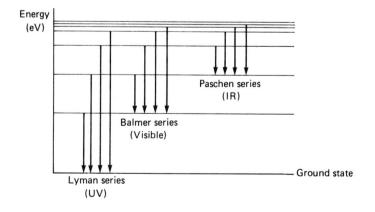

Figure 10.3

Paschen series (IR). These represent electrons moving to the ground and second excited energy states respectively.

The *greater* the transition, the *higher* is the frequency of radiation emitted. When materials are bombarded by high energy electrons, electrons close to the ground state are excited. The vacant places produced can be taken up by electrons from higher energy levels creating very high frequency radiation—*X-rays*.

Spectrometer

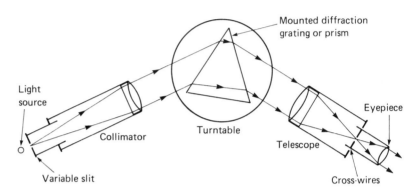

Figure 10.4

The spectrometer (Fig. 10.4) is an *optical device* used to investigate optical spectra, but angles of prisms and refractive indices can also be found by it.

Spectrometer parts and adjustments

A moveable slit inside a fixed *collimator* forms a parallel beam of light from the source. The width of the slit can be altered. The beam passes through a *prism* or *diffraction grating* set on a *turntable* and into a *telescope*. Both turntable and telescope are free to rotate about a circular *scale* showing angle of rotation.

The telescope is provided with a vernier scale for accurate measurement of angle and, like the collimator, contains an achromatic converging lens.

Before use the following *adjustments* are required.

(1) The *eyepiece* is adjusted until the cross-wires are clearly seen. A distant object is then viewed. The telescope length is altered by means of a thumb screw until the object appears to be at the same place as the cross-wires. This condition is called **non-parallax**. Thus, parallel rays from the object are focused at the cross-wires.

(2) The *collimator* is now adjusted so that parallel rays come from it by illuminating the slit with sodium light. With the turntable *clear*, the telescope is put opposite the collimator. As rays from it are not parallel the slit image is blurred but is made sharp by changing the slit position.

(3) If the slit image is off centre when the prism is replaced, the telescope axis of rotation is not parallel with the prism refracting edge. Consequently the table has to be *levelled*, which is done by adjusting three screws, A, B and C (Fig. 10.5).

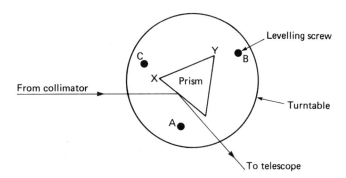

Figure 10.5

One of the prism faces is put perpendicular to the line AB and the collimator moved so that this face reflects light from it. The telescope is now moved until the reflection can be seen through it, and screw A is adjusted until the image is *centred*.

With the collimator and telescope *fixed* the turntable is rotated until prism face XY also gives a reflected image. Again this is centred, by adjusting screw C. The turntable has now been levelled.

Applications of the spectrometer

Finding the angle of a prism, *A*

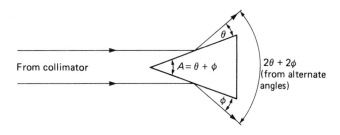

Figure 10.6

See Fig. 10.6. The prism is placed with its refracting edge towards the collimator. A reflected image is formed by two of the prism faces, and if the telescope is used to locate them, the angle between the two telescope positions gives *twice* the *prism angle A*.

Finding the minimum deviation, D_{min}

The minimum deviation for a prism is the minimum angle through which the emergent ray has been deviated from the original path of the incident ray.

The turntable, prism and collimator are set up as shown in Fig. 10.7. The slit image is then located by the telescope. Next the angle of incidence *i* (as shown) is *decreased* by slowly rotating the turntable. This causes the slit image to shift and so the telescope is also moved and the image kept centred. The telescope moves in the direction of the original path, but at a certain point moves away from it. This point is relocated by

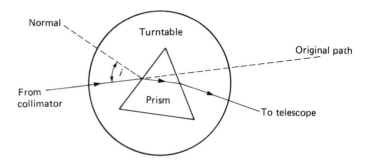

Figure 10.7

rotating the turntable back, and the angle between the position of the telescope at this point and the original path is D_{min}.

The telescope position is noted, the prism removed and the new telescope position which will give an image noted. The *difference* gives the angle D_{min}.

Refractive index of prism glass

If both A and D_{min} are found for a prism, the refractive index of the prism glass is given by

$$n = \frac{\sin\left(\dfrac{A + D_{min}}{2}\right)}{\sin\left(A/2\right)}$$

This method is very accurate and can also be used for *liquids* if they can be contained in a thin-walled, hollow, glass prism.

Wavelength measurement

For *reference*, the prism is put on the turntable in the minimum deviation position for sodium light. A helium discharge tube producing a line spectrum of several colours is now put in place of the sodium light source and each line viewed in turn by the telescope. Each time, the angle of deviation for the line D^* is noted. The procedure is repeated with a hydrogen discharge tube, a neon tube and alkali metal salts in bunsen flames as sources.

Wavelengths corresponding to the different deviations can then be found from *standard tables*. The resulting calibration curve of D^* versus wavelength λ (Fig. 10.8) will then give the wavelength for a line of an unknown source.

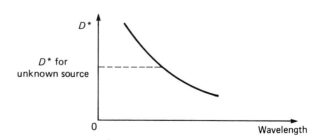

Figure 10.8

Example 10.1

The flame produced by a certain metal X has a colouration due to the transition of an electron from an excited state of $-12.34\,\text{eV}$ to its ground state of $-14.19\,\text{eV}$. Determine the wavelength of the light in the flame. (Assume Planck's constant $h = 6.6 \times 10^{-34}\,\text{J s}$ and $c = 3 \times 10^8\,\text{m s}^{-1}$.)

From Planck's theory, the quantum of radiation emitted is given by

$$E = hf = \frac{hc}{\lambda}$$

From the conservation of energy, this equals the amount of energy lost by the electron in its transition. Therefore

$$E_2 - E_1 = \frac{hc}{\lambda}$$

But $E_2 - E_1 = (14.19 - 12.34) \times 1.6 \times 10^{-19} = 2.96 \times 10^{-19}\,\text{J}$

and so

$$\lambda = \frac{6.6 \times 10^{-34} \times 3 \times 10^8}{2.96 \times 10^{-19}} = \underline{6.67 \times 10^{-7}\,\text{m}}$$

Example 10.2

From the spectrometer calibration curve in Fig. 10.9, estimate the minimum deviation that occurs in the spectrometer prism, when light of wavelength $0.535\,\mu\text{m}$ from a thallium flame is passed through it. If the refractive index of the prism glass is 1.51 for that light, determine (i) the wavelength of the thallium light in the prism glass, and (ii) the prism angle.

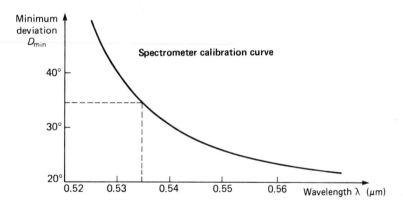

Figure 10.9

The minimum deviation as estimated from the graph is $\underline{34°}$.

(i)   For light passing from medium A to medium B, the refractive index is given by

$$_A n_B = \lambda_A / \lambda_B$$

Thus, $\lambda_B = \dfrac{0.535 \times 10^{-6}}{1.51} = \underline{0.354\,\mu\text{m}}$

(ii) $n = \dfrac{\sin[(A + D_{\text{min}})/2]}{\sin(A/2)} = \cos(D_{\text{min}}/2) + \cot(A/2)\sin(D_{\text{min}}/2)$

(from trigonometry), so

$$\cot(A/2) = \frac{n - \cos(D_{min}/2)}{\sin(D_{min}/2)}$$

Thus, $A = \underline{55.7^\circ}$

Interference in light

The superposition of light waves from two **coherent** sources (i.e. having constant phase difference) creates an interference pattern. This is *enhanced* if the sources are closer together or if the screen where the waves superimpose is further away. Such sources are only obtained from one point of a single light source since even when two points of a single source are used the changing phase difference destroys the pattern. The colours of thin oil films on wet roads and of soap bubbles are due to the interference of light.

Young's slits

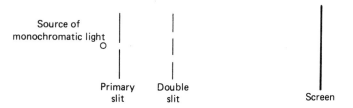

Figure 10.10

Monochromatic (one wavelength) light passes through a primary slit and illuminates two very narrow, parallel slits (Fig. 10.10). The *diffracted* waves from these two coherent sources then *interfere*, producing a series of dark and light fringes on a screen. The effect is destroyed by closing one of the parallel slits.

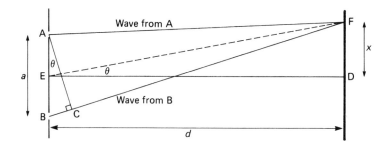

Figure 10.11

The waves reaching D (Fig. 10.11) are in phase, since AD and BD are equal. Thus, *constructive interference* takes place and a bright fringe is formed at D. If the path difference BC is a whole number of wavelengths $n\lambda$, a bright fringe is formed at F as well. Therefore

$$\boxed{\sin\theta = n\frac{\lambda}{a}}$$

where a is the slit separation.

As a is small compared to d, the angle FED is also θ and $\tan\theta = x/d$. Now θ is also very small and $\sin\theta = \tan\theta$ approximately. Therefore

$$n\frac{\lambda}{a} = \frac{x}{d}$$ (for the *n*th *bright* fringe)

Dark fringes are produced when the distance BC contains an odd number of half wavelengths and the two light waves arrive at F in antiphase, causing *destructive interference*. Thus,

$$\sin\theta = (n+\tfrac{1}{2})\frac{\lambda}{a}$$

and so

$$(n+\tfrac{1}{2})\frac{\lambda}{a} = \frac{x}{d}$$ (for a *dark* fringe)

Width of a fringe

If x and x' are the distances of the *n*th and $(n+1)$th bright fringes from the centre of the pattern respectively, then

$$x = \frac{n\lambda d}{a} \quad \text{and} \quad x' = (n+1)\frac{\lambda d}{a}$$

Therefore

$$x' - x = \frac{\lambda d}{a}$$

and so the fringe width is

$$w = \frac{\lambda d}{a}$$

Determination of the wavelength of light

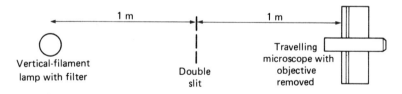

Figure 10.12

(1 m — Vertical-filament lamp with filter — Double slit — 1 m — Travelling microscope with objective removed)

A vertical-filament lamp fitted with a filter acts as a single slit and projects *monochromatic* light through the double slit and into a travelling microscope (Fig. 10.12). This is used as a screen since it can be moved from side to side and the displacements read from the vernier scale. An average *fringe separation w* is then found by measuring the distance across as many fringes as possible.

The double slit separation is found by using an ordinary travelling microscope. The wavelength can then be calculated with

$$\lambda = wa/d$$

The fringes in the experiment are formed along *hyperbolae* anywhere in the region of wave overlap. If white light is used, each colour gives a set of fringes. These overlap, although only the central one is white since there the path difference for all the colours is zero. The first bright fringe is coloured blue on the side nearest the centre and red on the far side because fringe width is proportional to wavelength.

If the slits are made *narrower*, the number of fringes *rises* but they become *dimmer*.

Interference due to thin films

A thin film of air is trapped in a wedge formed by two thin glass plates kept apart at one end by thin foil. It is illuminated by *monochromatic* light partially reflected off a glass sheet from an extended source. A microscope focused onto the wedge picks out bright and dark, equally spaced fringes (Fig. 10.13).

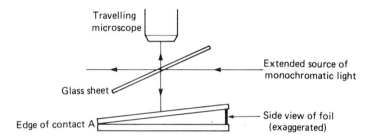

Figure 10.13

Thin film interference fringe pattern

The fringes formed by the air film wedge run *parallel* to the contact edge. *Partial* reflection of light occurs at the lower surface of the top glass plate making up the wedge, while the rest goes through the air film to be reflected at the top surface of the lower plate (Fig. 10.14).

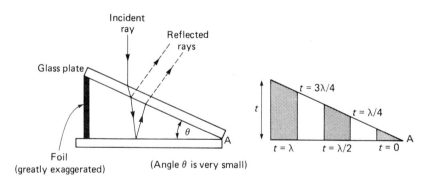

Figure 10.14

As both reflected waves originate from the same point X, they are *coherent* and *division of amplitude* arises. Interference is produced when an eye lens or eyepiece brings the waves together. Reflection of waves off a denser medium causes a phase change of 180° and this is equivalent to *adding* half a wavelength to the wave. So the *path difference* between the two reflected waves at X is

$$2t + \lambda/2$$

where t is the air film thickness. At point A, $t = 0$ and a dark fringe will occur since the path difference is half a wavelength.

For *bright* fringes,

$$\boxed{2t + \lambda/2 = n\lambda} \quad \text{(for the first bright fringe } n = 1, \text{ etc.)}$$

For *dark* fringes,

$$2t + \lambda/2 = (n + \tfrac{1}{2})\lambda$$

Therefore

$$\boxed{2t = n\lambda} \quad \text{(for the first dark fringe } n = 0, \text{ etc.)}$$

In order to observe the fringes which are located at the air film, the eye lens or microscope must be *focused there*. Fringes mark all points in the air film that have the *same* path difference, and should be straight bands if the optical contact along the edge where the plates meet is *perfect*. If not, they are distorted. However, they are not affected by reflections from the other plate surfaces since these have greater path differences and may not be in focus.

(1) Foil thickness may be determined by counting the number of bright fringes m between the foil and the contact edge.

If the foil is at a *bright* fringe, then

$$2t = m\lambda + \lambda/2$$

If the foil is at a *dark* fringe, then

$$t = m\lambda/2$$

where λ is the monochromatic light wavelength.

(2) As the wavelength of light increases, the film thickness at which fringes occur also increases, and the fringes get *further apart*.

Example 10.3

A Young's slits experiment is carried out with the slits placed 1.6 m away from a travelling microscope with its objective removed. When light of wavelength 0.63 μm is used to illuminate the slits, the separation of the seventh bright band on one side of the pattern centre and the centre itself is 24.8 mm. Find the slit separation.

Applying the expression

$$\frac{n\lambda}{a} = \frac{x}{d}$$

where n is the order of the fringe, then

$$a = \frac{nd\lambda}{x} = \frac{7 \times 1.5 \times 0.63 \times 10^{-6}}{24.8 \times 10^{-3}}$$

Therefore

$$a = \underline{2.7 \times 10^{-4}\,\text{m}}$$

Example 10.4
Using the set up in example 10.3, find (i) the distance between the centre of the pattern and the sixth dark fringe to one side of the centre, and (ii) the fringe width.

(i)   For the nth dark fringe to one side of the pattern centre,

$$\frac{(n+1/2)\lambda}{a} = \frac{x}{d}$$

where n is the dark fringe order.
Therefore

$$x = \frac{d(n+1/2)\lambda}{a} = \frac{1.5(6+1/2) \times 0.63 \times 10^{-6}}{2.7 \times 10^{-4}}$$

Thus,
$$x = 2.28 \times 10^{-2}\,\text{m}$$

(ii)   The fringe width is given by

$$w = \frac{\lambda d}{a} = \frac{0.63 \times 10^{-6} \times 1.6}{2.7 \times 10^{-4}} = 3.7 \times 10^{-3}\,\text{m}$$

Example 10.5
A thin wedge-shaped film of air is formed by two thin glass plates placed on top of each other with a piece of tissue separating them at one end. If monochromatic light of wavelength 0.53 μm is used to illuminate the air wedge and the distance between the contact edge and the edge of the tissue is 64 mm, find the tissue thickness which produces 33 bright fringes along 4 mm of the wedge. What would be the effect of increasing tissue thickness?

The number of bright fringes between the contact edge and the edge of the tissue is given by

$$m = \frac{64 \times 33}{4} = 528 \text{ bright fringes}$$

The thickness t of the tissue is given by
$$2t = m\lambda + \lambda/2$$

Thus,

$$t = \frac{\lambda}{2}(m + \tfrac{1}{2})$$

Therefore

$$t = \frac{0.53 \times 10^{-6}}{2}(528 + \tfrac{1}{2}) = 1.4 \times 10^{-4}\,\text{m}$$

Assuming the tissue to be at a dark fringe,

$$t = \frac{528 \times 0.53 \times 10^{-6}}{2} = 1.4 \times 10^{-4}\,\text{m}\quad\text{(also)}$$

As the thickness of the tissue increases, the number of fringes appearing per unit length of the air wedge will increase also. (Decreasing the wavelength for the same tissue thickness would have the same effect.)

Coloured films

Thin glass, oil or soap films also exhibit interference. When focused by a lens the reflected rays interfere. There is a *path difference* due to refraction and to a phase change off the top film surface, which varies with the angle of refraction. It thus also varies with the angle of *incidence* and the angle of reflection at the top film surface (Fig. 10.15).

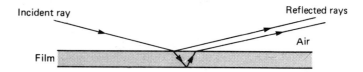

Figure 10.15

When white light from a source such as a cloud illuminates a film, changing the position of the *lens* changes the angle of reflection and the wavelength of the light seen interfering. This results in seeing *different* coloured bands.

A vertically placed, thin film of soap slowly 'flows' downwards, creating a vertical wedge of soap film. If this is illuminated by white light, horizontal fringes occur.

Newton's rings

Coherent sources are again formed here by *division* of *amplitude*. Monochromatic light is reflected from a glass plate and is incident *normally* upon a plano-convex lens of long focal length, resting on another glass plate (Fig. 10.16). The thickness of the *air film* in a horizontal circle about the point A is the same. Interference fringes viewed through the travelling microscope focused onto the air film are due to reflections off the lower lens surface and the upper glass plate surface.

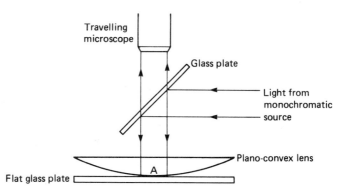

Figure 10.16

They occur as *rings* centred at point A, whose separation decreases outwards from A, where there is perfect optical contact and a dark spot is formed. Although the air film thickness at A is zero, there is still a phase change due to reflection.

(1) *Distortions* in Newton's rings show that a lens requires further grinding or that a surface is not optically flat.

(2)   When *white* light is used, coloured rings are formed, but the range of colours decreases outwards until white light is produced by overlapping colours. Sodium light gives rings of varying clarity because it consists of light of two wavelengths.

Blooming

To reduce reflected light from optical surfaces, a thin transparent coating of magnesium fluoride is applied. Destructive interference is achieved when this layer is a quarter of a wavelength thick, as air–coating and coating–glass boundaries cause a phase change of 180° on reflection. Such coating is called **blooming**.

Diffraction of light

The superposition of waves from *coherent* sources on a wavefront after that wavefront has been *distorted* by an obstacle or aperture is called **diffraction**, and a diffraction pattern is formed. It is most evident when the linear dimensions of the obstacle or aperture *approach* the wavelength of the waves used.

Single and double slit diffraction

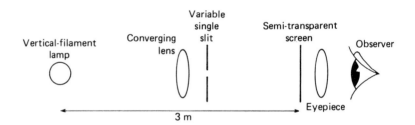

Figure 10.17

See Fig. 10.17. With a wide *single* slit separation, the converging lens is adjusted to focus an image of the vertical filament onto the screen and a diffraction pattern is seen as the slit is narrowed. For *white* light a central white band with coloured bands (maxima), either side, separated by dark bands is seen. Filtered light gives bands of a *certain* colour only, whose separation decreases with wavelength used—e.g. red bands are further apart than blue bands. The number of bands from the centre is called the **order** of the band.

As a screen can be placed at any distance from the slit, the separation of bands is expressed in terms of *angle* with respect to the central bright band rather than in metres. This central band is twice as wide as any other. A plot of *relative band intensity* versus angular width is called a **Fraunhofer intensity pattern**.

For a *double* slit the bands become more *distinct* and for many slits the effect is even greater. The Fraunhofer patterns for them show two components—an overall **diffraction envelope** due to the interference occurring between secondary wavelets from a single slit, and an **interference component** due to interference between secondary wavelets from different slits (Fig. 10.18).

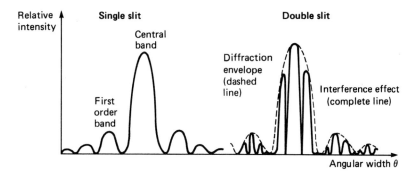

Figure 10.18

Transmission diffraction grating

A transmission grating allows light to pass between many fine, equally spaced, parallel lines ruled onto a glass plate. They can be used for *measuring wavelength* and give very *distinct spectra*.

A fine diffraction grating may have 500 lines or more per mm and produces a series of bands when illuminated by while light as shown in Fig. 10.19.

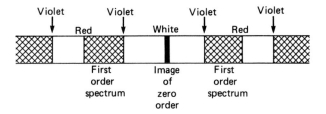

Figure 10.19

Effect of a transmission grating on white light

In comparison with spectra produced by prisms, it can be seen that the red is *deviated more* than the violet with the grating, and dispersion increases as spectral order increases. The intensity of grating spectra is *less* than that of prism spectra as radiant energy is distributed over a number of peaks in the Fraunhofer pattern. Also wavelengths in the latter are compressed at the red end while in the former they are evenly distributed.

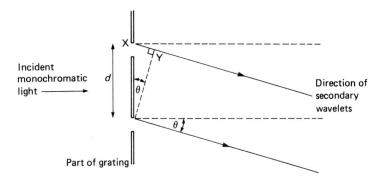

Figure 10.20

For all pairs of wavelets along the grating in the direction θ to the line of the incident beam, the path difference XY will be $d \sin \theta$ (Fig. 10.20). So, for a *maximum*,

$$d \sin \theta = n\lambda$$

where d is the spacing of the grating and n is the order of the image or spectrum.

When the path difference of all the diffracted secondary wavelets is zero, $d \sin \theta = 0$ and the **zero order image** is formed. The maximum order spectrum possible is

$$\frac{d}{\lambda} \sin 90° = \frac{d}{\lambda} \qquad \text{(to the nearest and lowest integer)}$$

As seen from the Fraunhofer intensity pattern for many slits (Fig. 10.21), the intensity is zero when the intensity due to the diffraction component is zero. If a certain order image should appear at that position, it will be absent.

Fraunhofer pattern for multiple slits

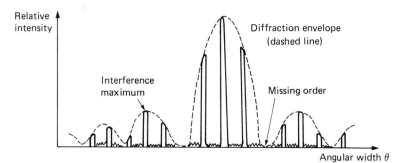

Figure 10.21

Angular width θ

Wavelength measurement

A *transmission* diffraction grating mounted onto a spectrometer, which has had telescope, collimator and cross-wires adjustments applied to it, can be used to find the wavelength of monochromatic light. The telescope is put perpendicular to the collimator and the grating is mounted on the spectrometer table so that it is *perpendicular* to the line joining two levelling screws. The table is moved until a reflection of the collimator slit, off the grating, is observed through the telescope (Fig. 10.22(a)). Adjustment of either of the two screws mentioned above will *centre* the image, bringing the telescope axis of rotation *parallel* to the plane of the grating.

Normal incidence of light is obtained by moving the table through 45° and the telescope turned until the *first order* image is observed. Adjustment of the *third* levelling screw will centre the image and the scale reading is noted (Fig. 10.22(b)). The first order image on the other side is next found and the angle 2θ between the two positions recorded. If d is the grating spacing and n is the order of the image (i.e. one), then

$$d \sin \theta = n\lambda$$

so $\boxed{\lambda = d \sin \theta}$

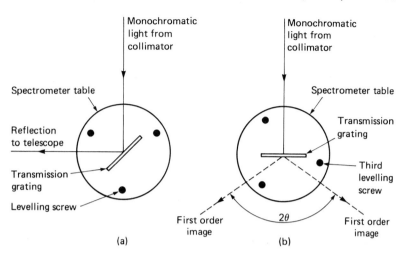

Figure 10.22

(a)   (b)

Example 10.6

Two wavelengths of light are incident simultaneously on a transmission diffraction grating, which has 3000 lines per cm. The fifth order image due to one wavelength occurs at the same place as the sixth order image of the other, in the diffraction pattern. If one wavelength is greater than the other by 0.11 μm, find (i) the values of the two wavelengths and (ii) the angular displacement away from the centre of the diffraction pattern at which the coincidence occurs.

(i)   Let the two wavelengths be λ and λ'. Then

$$\lambda' = \lambda + (0.11 \times 10^{-6})$$

Applying the expression

$$d \sin \theta = n\lambda$$

then $n_1\lambda = d \sin \theta$ and $n_2(\lambda + 0.11 \times 10^{-6}) = d \sin \theta$

Therefore

$$n_1\lambda = n_2(\lambda + 0.11 \times 10^{-6})$$

Thus, $\lambda = \dfrac{n_2 \times 0.11 \times 10^{-6}}{n_1 - n_2} = \dfrac{5 \times 0.11 \times 10^{-6}}{6 - 5}$

Therefore

$$\lambda = 5.5 \times 10^{-7}\ \text{m} \quad \text{and so} \quad \lambda' = 6.6 \times 10^{-7}\ \text{m}$$

(ii)   Now,

$$\sin \theta = n_1\lambda/d$$

where d is the grating spacing. But

$$d = 1/(3000 \times 100) = 3.33 \times 10^{-6}\ \text{m}$$

Therefore

$$\sin \theta = \frac{6 \times 5.5 \times 10^{-7}}{3.33 \times 10^{-6}} = 0.991$$

So $\theta = 82.3°$

Example 10.7
Monochromatic light is normally incident on a transmission diffraction grating which has 5000 lines per cm. The angle of diffraction for the first order image is 23°. When the grating is placed into a beaker of methanol and once again illuminated with the same light, the angle of diffraction is reduced. If the refractive index for methanol is 1.36, determine (i) the wavelength of the light in methanol, and (ii) the new angle of diffraction for the first order image.

(i) Applying the expressions

$$d \sin \theta = n\lambda$$

and $d = 1/N$ (where N is the number of lines per metre)

also $n = 1$ for the first order image, then

$$\lambda = \sin 23° / N$$

But $_1 n_2 = \lambda_1 / \lambda_2$

and thus,

$$\lambda_{\text{methanol}} = \frac{\sin 23°}{_1 n_2 N} = \frac{0.3907}{1.36 \times 5 \times 10^5}$$

Hence

$$\lambda_{\text{methanol}} = \underline{5.8 \times 10^{-7} \text{ m}}$$

(ii) If the new angle of diffraction is θ', then

$$\sin \theta' = N\lambda_{\text{methanol}} = \underline{5 \times 10^5 \times 5.8 \times 10^{-7}}$$

Therefore

$$\theta' = \underline{16.9°}$$

Electromagnetic spectrum

Radiation	Frequency (Hz)	Wavelength (m)	Method of production
Long electric waves	$1 - 10^4$	$3 \times 10^8 - 3 \times 10^4$	Rotating coil within a magnetic field
Radio waves	$1 \times 10^4 - 1 \times 10^9$	$3 \times 10^4 - 3 \times 10^{-1}$	Electrical oscillations in inductive and capacitative circuitry
Microwaves	$1 \times 10^9 - 3 \times 10^{11}$	$3 \times 10^{-1} - 1 \times 10^{-3}$	Electrical oscillations in inductive and capacitative circuitry
Infrared			
Far	$3 \times 10^{11} - 6 \times 10^{12}$	$1 \times 10^{-3} - 5 \times 10^{-5}$	Produced from hot bodies such as
Near	$6 \times 10^{12} - 4 \times 10^{14}$	$5 \times 10^{-5} - 7.8 \times 10^{-7}$	stars and fires
Visible	$4 \times 10^{14} - 8 \times 10^{14}$	$7.8 \times 10^{-7} - 3.8 \times 10^{-7}$	Produced by hot solids, liquids, gases or by ionised gases at low pressure
Ultraviolet	$8 \times 10^{14} - 6 \times 10^{16}$	$3.8 \times 10^{-7} - 5 \times 10^{-9}$	Produced by arc and spark discharges, mercury vapour lamps and very hot bodies
X-rays	$6 \times 10^{16} - 1 \times 10^{23}$	$5 \times 10^{-9} - 3 \times 10^{-15}$	Produced by bombardment of heavy metal targets with fast electrons
Gamma rays	$1 \times 10^{19} - 1 \times 10^{21}$	$3 \times 10^{-11} - 3 \times 10^{-13}$	Produced as a result of radioactive decay
	$1 \times 10^{21} - 1 \times 10^{24}$	$3 \times 10^{-13} - 3 \times 10^{-16}$	Produced as a result of cosmic ray interactions with nuclei of atoms

An **electromagnetic wave** consists of oscillating electric and magnetic fields *perpendicular* to each other. All such waves are grouped into a continuous frequency range called the **electromagnetic spectrum**. There are no distinct boundaries for the regions on it and because of the different methods of production some *overlap* does occur.

Wavelengths for the waves can be obtained from the wave equation, but these are *dependent* upon the *medium* of transmission, whereas frequency stays the same no matter what medium is used. The high frequency waves show properties which favour the corpuscular theory which suggests that radiation has a particulate nature. The *quanta* here, given by *hf* where *h* is *Planck's constant*, are very 'distinct'. Low frequency radiations exhibit wave-like properties. Their quanta are less 'distinct'.

Electromagnetic waves

The various waves on the electromagnetic spectrum share a number of properties.

(1) They can all propagate through a *vacuum* and all travel with the *same speed*, which is given the symbol *c*.

$$c = 3 \times 10^8 \text{ m s}^{-1}$$

(2) They all suffer *reflection* on striking a body, although partial absorption does occur. The absorption leads to an increase in internal energy of the body.

(3) They can all be *refracted* or *diffracted* and all exhibit *interference*.

(4) They are all *transverse* waves and can all be *polarised* since this is a characteristic property of such waves.

Radio

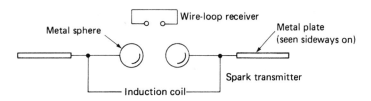

Figure 10.23

Radio waves can be obtained using the simple apparatus shown in Fig. 10.23. Pulses of current from the induction coil charge up the plates of the transmitter until a discharge occurs between the spheres. This produces an oscillatory current in the transmitter circuit, and the resulting radio waves are detected by sparks across a gap in a wire loop placed some distance away. Radio waves can be *refracted* in a large *asphalt* prism and are *reflected* off *metal* plates.

Electronic oscillators containing transistors are now used to transmit radio waves via aerials. Electrons in an aerial are caused to accelerate up and down it, resulting in an electromagnetic wave whose electric field is in the plane of the aerial. The waves are **plane polarised**.

Transmission of radio waves across the Earth is achieved by reflecting them off a layer of ionised gas called the **ionosphere**. It lies about 50 km above the Earth's surface and waves of different frequency are reflected from it at different levels. Waves of certain frequency, however, pass through it and require communications satellites for their transmission.

Radio applications

Radio telephony, telegraphy and television are examples of radio applications. Radio control uses 'frames' made up of a series of pulses which are converted to mechanical signals inside a device such as an industrial robot, model plane, or bomb-investigation machine. Radio-collars are used to track the movements of animals, and sensors fitted with transmitters can relay information about enemy troop movements.

Radio telescopes are very sensitive radio receivers. Large dishes gather signals emitted by sources in space. These are amplified and can be displayed by a computer in the form of a radio 'map'. Even larger dishes enable finer detail to be examined—the *resolution* improves. For even better results, radio telescopes are linked electronically and the resulting *interference patterns* investigated. (Gases and matter trapped in the magnetic fields of stars emit radio waves.)

Microwaves

These are high frequency radio waves. They basically cover the range used in radar—radio detection and ranging. Radio pulses are targeted at bodies such as ships or planes and the reflected signals detected. The direction of the body is given by the detecting aerial, and the range is obtained from the time interval between pulse emission and arrival of reflection. Systems also use the *Doppler effect* to determine the speed of the body. Another use is the microwave cooker, while in schools the *3 cm apparatus* is used to demonstrate wave properties.

Infrared

All bodies emit infrared over a continuous frequency range, although the amount emitted at any particular frequency depends upon the *temperature* of the body and the type of surface it has. As the temperature increases the intensity of emitted radiation also rises and higher frequency radiation becomes *more evident*. For example the colour changes of iron when it is heated.

Infrared can be detected by a blackened thermometer when it is placed beyond the red end of an optical spectrum produced by a prism. Other detectors include *phototransistors*, infrared film, *thermopiles* and bolometers.

The increase of *internal energy* when a body absorbs infrared leads to an increase in temperature and a feeling of warmth when absorbed by the skin; hence the use of infrared lamps in physiotherapy and paint drying. It is also used for detection at night by the army, or biologists investigating the activities of nocturnal animals. Infrared is used to check power cables. If a cable is not conducting properly, it will appear darker than cables which are, in an infrared camera. In the world of art, pictures can be authenticated by infrared.

When sunlight is absorbed by plants in a greenhouse, some of the energy is *re-emitted* in the form of low frequency infrared which is unable to escape through the glass and so causes warmth in the greenhouse. A similar effect occurs in the atmosphere, where carbon dioxide layers act like the glass. The effect is most evident on the planet Venus.

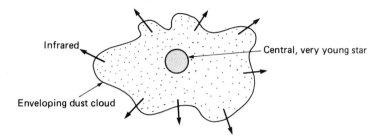

Infrared

Central, very young star

Enveloping dust cloud

Figure 10.24

The use of infrared detectors in *astronomy* has enabled astrophysicists to investigate the early stages of stellar evolution. *Dust clouds* are caused to *emit infrared* by the heating action of forming stars within them (Fig. 10.24). Organic molecules also give characteristic infrared spectra which has led to the identification of a whole group of 'space chemicals'.

Visible radiation

This is generated by hot bodies or ionised gases at low pressures. The visible region of the electromagnetic spectrum is comparatively small but it is the radiation to which human eyes are sensitive. There are many applications—warning mechanisms e.g. lighthouses, energy source, communications, and lasers.

Ultraviolet

This is radiation beyond the violet end of the visible spectrum and is able to initiate chemical changes. Ultraviolet can be used to demonstrate the *photoelectric effect* and to cause certain substances to glow—called *fluorescence*. This occurs with teeth, finger nails, egg shells, paraffin oil and materials treated with modern washing powders.

Fluorescence is the absorption of radiation and the emission of lower frequency radiation, ceasing when the radiation source is removed. In *phosphorescence*, the emission continues even when the radiation source is removed.

A mercury vapour lamp produces ultraviolet, but is fitted with quartz windows since glass absorbs ultraviolet. The lamp should never be looked at directly, since it can cause eye damage, although UV does lead to vitamin formation in the skin. It tans skin but over-exposure can be dangerous. An ozone layer in the atmosphere *filters out* most of the harmful ultraviolet from the Sun but it is in danger of disintegration because of pollution.

'Strip lighting' consists of mercury vapour filled tubes which are coated on the inside with fluorescent powder. This glows when ultraviolet produced by an electrical discharge through the vapour strikes it.

X-rays

X-rays expose photographic film and can be used to investigate bone structure. They are produced by *X-ray tubes* (Fig. 10.25). These consist of a *concave cathode* which focuses electrons emitted by a tungsten filament connected to it. The electrons pass through an evacuated chamber to a tungsten target embedded in a copper block. Most of the electron energy is converted to heat and so the copper block is *water cooled*. The block acts as *an anode* and the accelerating p.d. across it and the cathode is about 100 kV. X-rays from the target penetrate a window in lead shielding which surrounds the tube. A thickness of about 1 mm of lead is needed.

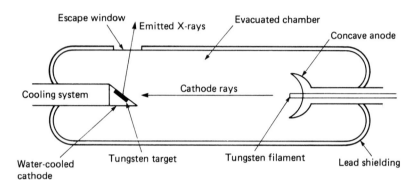

Figure 10.25

The *number of electrons* striking the tungsten target determines the *intensity* of X-rays emitted and can be varied by altering the current in the filament. X-ray frequency is determined by the *accelerating p.d.*

X-rays exhibit all the properties common to electromagnetic radiation and the properties of ultraviolet, i.e. the photoelectric effect and the ability to cause fluorescence. In addition they are able to ionise gases.

X-ray diffraction
This can be achieved by passing a collimated beam of X-rays through a crystal onto a photographic plate. The atomic lattice of the crystal acts like a *3-d* transmission *diffraction grating*. Since the atomic spacing is of the order of the X-ray wavelength, a diffraction pattern is observed. This consists of an arrangement of dots around a central spot—formed by radiation going straight through the crystal—and is called a *Laue pattern*.

In the Bragg method, the atoms scatter the incident radiation. The scattered X-rays *interfere* and weak reinforcement occurs when the angle of incidence equals the angle of scatter with respect to the normal to the atomic plane.

X-ray spectra
Any particular target emits X-rays with a range of frequencies (Fig. 10.26). Such X-ray spectra have *two components*. The **continuous component** is produced by *all targets* and has a sharp minimum wavelength cut-off. Electrons causing X-ray emission strike the target with energy eV, where e is the electronic charge and V is the accelerating p.d. At best all of the energy can be given to an X-ray photon. If f is the

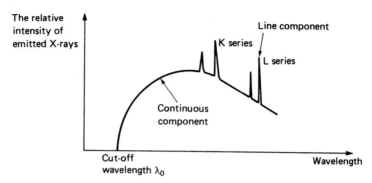

The relative intensity of emitted X-rays

K series

Line component

L series

Continuous component

Cut-off wavelength λ_0

Wavelength

Figure 10.26

maximum X-ray frequency, then

$$eV = hf$$

Now

$$hf = \frac{hc}{\lambda_0}$$

and so

$$\lambda_0 = \frac{hc}{eV}$$

Given values of c, e, V and of λ_0—the minimum wavelength cut-off which is determined from spectrometer measurements—the value of Planck's constant h can be obtained.

The **line component** superimposed onto the continuous spectrum is characteristic of the target material. The groups or series of lines are called K, L and M, in order of increasing wavelength. For low nucleon number elements, only the K series is observed, but with increasing nucleon number the wavelengths at which the series occur get shorter and eventually the L and M series appear.

Uses of X-rays include the treatment of cancer with hard (high frequency) X-rays and the detection of cracks and defects in machinery normally inaccessible. X-ray detectors are also used in astronomy. (Theory predicts black holes to be regions of X-ray emission.)

Gamma rays

Essentially these are much like X-rays but differ in the important respect that they are produced *from the nucleus*, whereas X-rays are produced from electron transitions close to it.

Polarised light

The polarisation of light waves is referred to the plane of the oscillating *electric field*. This, in the main, is responsible for the interaction between light and matter. For example, if light is horizontally polarised, the electric field is in the horizontal plane. The particular arrangement of crystals in a polaroid lens only permits the transmission of light waves with electric fields in a particular plane—the plane of polarisation. Absorption of those with electric fields perpendicular to that plane occurs.

When light is *partially reflected* at the boundary of certain media, e.g. air–glass for defined angle of incidence, the reflected ray becomes totally plane polarised. The angle of incidence is then called the **polarising angle**.

Another method of polarisation is the effect of scattering unpolarised light in a *suspension* of milk in water. This can be confirmed by viewing the scattered light through polaroid filters. Calcite crystals produce an effect called *double* refraction where incident unpolarised light is split into two plane polarised refracted rays. The planes of polarisation of the two rays are perpendicular to each other.

Shiny or smooth surfaces such as water reflect light which is, at the least, *partially* polarised. Polaroid glasses or camera lenses reduce this so-called *glare*.

Optical activity

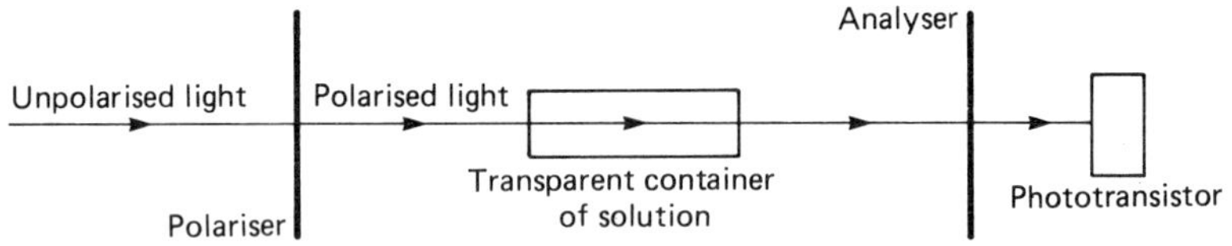

Figure 10.27

This is the ability of some crystals and liquids to *rotate* the plane of polarisation of polarised light passing through them. The extent of rotation depends upon a number of factors when a solution is being investigated. These include the wavelength of light and the concentration and the temperature of the solution. A *polarimeter* is a device for measuring the angle of rotation and consists of

(1)   a source of monochromatic polarised light—the *polariser* (e.g. polaroid filter),
(2)   a tube containing the solution under investigation,
(3)   a rotatable *analyser* (e.g. a polaroid filter.)

With the tube empty, the analyser is adjusted until the polarised light is just extinguished (Fig. 10.27). The tube is then filled with the solution and the analyser rotated until the light is extinguished again. The angle of rotation can then be found and the concentration of the solution calculated.

Photoelasticity

When white light passes through glass or some plastics under stress, *coloured fringes* are observed emanating from the stress areas if the body is viewed through crossed polaroids, i.e. polaroids acting perpendicular to each other. Plastic models of complex engineering designs can thus be *stress analysed*. Glass or plastic sheeting which has undergone some compression in forming appears to have coloured bands when viewed through polaroids.

Refracting astronomical telescope

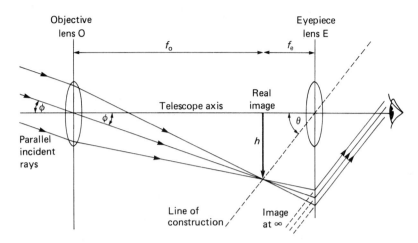

Figure 10.28

In this device (Fig. 10.28), rays of light from an object so distant that the rays are essentially parallel pass into a long focal length converging lens called the **objective**. It produces a real, inverted but diminished image which is in turn formed into a magnified, virtual image at infinity by a converging lens of *short focal length*. This is called the **eyepiece** and the telescope is said to be in **normal adjustment**.

In this adjustment, the distance between the real image and the eyepiece must be the eyepiece focal length. The distance between the real image and the objective is the objective focal length.

Angular magnification (magnifying power)

The expression for linear magnification is not applicable to the astronomical telescope as the image distance is infinity. Instead *angular* magnification, which can be generally applied to optical instruments, is used. It is defined as the ratio of the angle subtended at the eye by the image formed when the instrument is used, θ, to the angle subtended at the unaided eye, ϕ, for a set object distance.

$$M = \theta/\phi$$

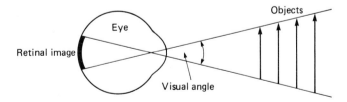

Figure 10.29

The angles θ and ϕ are called the *visual angles* (Fig. 10.29) and different sized objects will give the same visual angle. So angular magnification is in fact a comparison of apparent sizes of image, for a set object distance, whereas linear magnification is a ratio of true image and object sizes. If h is the real image height and f_e and f_o are the eyepiece and objective focal lengths respectively, then

$$\theta = h/f_e$$

and since the lens separation is much smaller than the object distance

$$\phi = h/f_o$$

for the *telescope* in *normal* adjustment.

Telescope angular magnification and the near point

Applying the expressions for the visual angles determined for the refracting astronomical telescope in *normal* adjustment, its angular magnification is given by

$$\boxed{M = \frac{h/f_e}{h/f_o} = \frac{f_o}{f_e}}\quad \text{(in normal adjustment)}$$

The **near point** is the position at the least distance from the eye where an object can be seen clearly, without strain. The distance to the near point is normally taken as 0.25 m. For the refracting astronomical telescope to form an image at the near point, the first real image must be closer to the eyepiece than f_e. The angular magnification is then just greater than f_o/f_e and the telescope is *out of* normal adjustment.

In normal adjustment, however, the eye is *unaccommodated* and so is relaxed, and this type of telescope setting is preferred.

Eye ring (Ramsden disc)

The best position for the eye when using an optical instrument is where *all* the light that has entered the instrument passes into the eye. In the case of the refracting telescope, all the rays passing through the objective pass through the image of the objective formed by the eyepiece. This image is termed the **eye ring**.

From the lens formula, the image distance v of the eye ring for the telescope in normal adjustment is given by

$$\frac{1}{v} = \frac{1}{f_e} - \frac{1}{u}$$

For normal adjustment the object distance $u = (f_o + f_e)$. Therefore

$$v = \frac{f_e}{f_o}(f_e + f_o)$$

From the concept of *linear* magnification, applied to the image of the *objective* in the eyepiece,

$$m = \frac{\text{diameter of eye ring}}{\text{diameter of objective}} = \frac{v}{u}$$

Thus,

$$m = \frac{f_e(f_e + f_o)/f_o}{f_e + f_o}$$

and

$$m = \frac{f_e}{f_o}$$

Now,

$$M = \frac{f_o}{f_e}$$

and so

$$M = \frac{\text{diameter of objective}}{\text{diameter of eye ring}}$$

The diameter of the eye ring can be found by placing a suitably illuminated object in front of the objective and eyepiece, in normal adjustment, and focusing the eye ring onto a screen.

Resolving power

The resolving power of an optical instrument is the extent to which it can separate objects that are *close together*.

Lenses, in the refracting astronomical telescope for example, act like circular *apertures* and produce *diffraction pattern* images of very small objects. The extent of overlap of these images determines the resolving power of the instrument.

Rayleigh's criterion

Two very small objects, close together, viewed by a refracting astronomical telescope produce the *Fraunhofer* intensity patterns shown in Fig. 10.30.

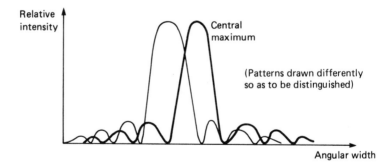

Figure 10.30

Rayleigh's criterion proposes that two *point objects* are just resolved when the *first order minimum* of one diffraction pattern is coincident with the central *maximum* of the other pattern, as shown in Fig. 10.30. The angular separation between the two central maxima for a circular aperture is given approximately by the ratio of the wavelength of light used to the diameter of the aperture.

Thus, increasing a telescope's objective diameter increases its resolving power, as one would like to resolve objects with as *small* an angular separation as possible. Hence radio telescopes require much larger dishes than lenses in refracting telescopes to obtain equivalent resolutions. (They are using *longer* wavelength radiation.)

An increase in objective diameter also allows more light to enter the telescope. Stars appear in the night sky as points and therefore their images as formed by astronomical telescopes will be point images. These get brighter as the objective diameter increases and stars which were previously too faint to be seen can be observed. The background to the stars is not a point source. It is called an **extended source** and, although more light is gathered from it when the diameter of the objective increases, there is also a corresponding increase in the area of the image. As a result the brightness of the background is unaltered.

Example 10.8   A refracting astronomical telescope consists of two lenses: an objective of focal length 1.5 m in normal adjustment with an eyepiece of focal length 0.04 m. Find (i) the angular magnification of the telescope and (ii) the distance of the eye ring from the eyepiece.

(i)   The angular magnification of the telescope is given by

$$M = \frac{f_o}{f_e} = \frac{1.5}{0.04} = \underline{37.5} \quad \text{(for normal adjustment)}$$

(ii)   The eye ring is the image of the objective formed by the eyepiece of the telescope, through which all the light entering the instrument passes. For normal adjustment, its distance v from the objective is given by

$$v = \frac{f_e(f_e + f_o)}{f_o} = \frac{0.04(0.04 + 1.5)}{1.5} \quad .$$

Thus,

$$v = \underline{0.041 \text{ m}}$$

11
Semiconductors

Introduction

The topic of semiconductors considered in this chapter is relatively new to Physics A level syllabuses. It begins therefore with two alternative, but equally valid, explanations of what a semiconductor is. These are supplemented with a section dealing with the effect of impurities on semiconductor properties and the explanation of majority and minority carriers.

This is followed by a discussion of the diode—its characteristics and application to the rectification of alternating current. The subsequent sections concern the bipolar transistor and its characteristics. This is a good follow-on from the diode, as the npn transistor can be thought of as two back-to-back junction diodes (npn-type as distinct from the pnp-type of transistor).

The last part of the chapter deals with the two major applications of the transistor. These are amplification and its use in logic circuits. A discussion of the basic type of amplifier is followed by sections on the various kinds of logic gate. These logic gates can be combined and are used widely in computer circuitry.

Revision targets

The student should be able to:

(1) Define the term semiconductor.
(2) Discuss how semiconductors can be distinguished from conductors and insulators.
(3) Explain what is meant by a 'hole' and define the term intrinsic semiconductor.
(4) Appreciate the significance of lattice dislocations and impurities upon the properties of a semiconductor.
(5) Define and distinguish between a donor impurity and an acceptor impurity.
(6) Distinguish between n-type and p-type semiconductors with reference to charge carriers.
(7) Describe the working of a p–n junction diode and in particular the meaning of reverse and forward bias.
(8) Sketch the characteristics for a p–n junction diode and explain how it can be applied to the rectification of a.c.
(9) Describe the basic construction of the npn transistor, and the relationship between emitter, base and collector currents.
(10) Describe the experimental determination of the common emitter mode transistor characteristics.

(11)   Sketch and discuss the main features of the transistor characteristics when the transistor is in the common emitter mode and compare them to the characteristics in the common base mode.

(12)   Give expressions for the hybrid parameters h_{fe} and h_{ie} and explain how they are related to the transistor characteristics.

(13)   Define and calculate the voltage gain of a basic amplifier.

(14)   Discuss the use of the transistor as a switch.

(15)   Discuss the major types of logic gate and summarise their functions in terms of truth tables.

(16)   Appreciate the combination of logic gates and their usefulness in computer circuitry.

What is a semiconductor?

A *semiconductor* is a material whose resistance decreases with increasing temperature and the addition of impurities.

Energy band theory

The properties of matter are not only dependent upon the arrangement of atoms within it, but also upon the nature of the atoms themselves. Electrons in an atom occupy **energy levels** and these are depicted on energy level diagrams as horizontal lines. The electrons are not allowed to exist between these 'lines'.

When atoms bind together to form a crystal, the upper energy levels (of the so-called *valence electrons*) are altered. This is because the valence electrons are no longer attracted by a *single nucleus*, but are also attracted by neighbouring ones. Thus, the energy needed to take an electron from a nucleus to an adjacent one is reduced.

The result is an increase in the number of *energy levels*.

Regarding the crystal as a whole, the many energy levels produced form *distinct energy bands*. At low temperatures and for an isolated crystal, the electrons fill, or fill most of, the lowest energy band called the **valence band**.

In order to set up an *electric current*, electrons must either be in a partly filled valence band or be promoted by the addition of energy to the band above, called the **conduction band**. Application of an electric field across the crystal then accelerates the electrons giving them energy. These electrons are termed *free electrons*.

Metals

In metals the energy bands *overlap* (Fig. 11.1). The valence band is mostly or completely filled. This enables metals to conduct electricity (and heat), at all temperatures. Although there is a small increase in the number of

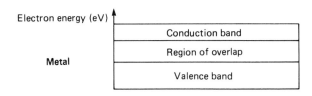

Figure 11.1

free electrons with temperature, their interaction with electron 'clouds' of the metal atoms impedes the flow of charge. Resistance *increases* with temperature.

Insulators

For insulators the valence band is wholly filled and an energy gap of about 5 eV separates it from the conduction band (Fig. 11.2). The gap is called the **forbidden energy gap**, where electrons are not allowed to stay. If a strong enough electric field is applied across the insulator, electrons are able to bridge the gap. Breakdown of the insulator is said to occur, and it will conduct electricity. Alternatively, energy can be supplied in the form of heat and so conductivity *improves* with increase in temperature.

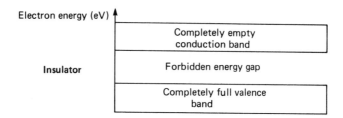

Figure 11.2

Semiconductors

For this class of material the valence band is completely filled (Fig. 11.3), but the forbidden energy gap is *much smaller* than for an insulator (about 1 eV). Even at room temperature a few electrons are able to bridge this gap into the conduction band. Conductivity *improves* with temperature rise.

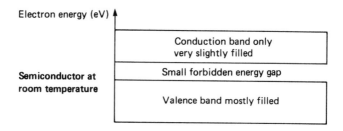

Figure 11.3

Holes

The electrons which transfer to the conduction band leave behind regions of *effectively positive charge* called **holes**. On the application of an electric field, the conduction electrons are accelerated in one direction. At the same time electrons in the valence band are able to take up the positions of holes and in turn they leave other positive holes behind. The net effect can be expressed as a current made up of free *electrons* moving in *one direction* and positive *holes* in the *opposite direction* (Fig. 11.4).

The holes *act like* positive electrons. Some of the conduction electrons fall back to *recombine* with holes in the valence band. The rate at which

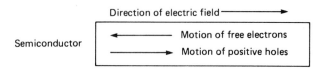

Figure 11.4

this happens, however, equals that of the appearance of conduction electrons. Pure semiconductors such as *germanium* or *silicon* contain equal numbers of holes and electrons. They are known as **intrinsic semiconductors.**

The above notes are an attempt to explain the nature of a semiconductor in terms of a *simplified* energy band theory. An alternative approach is as follows.

Covalent bonding theory

The atoms in a semiconductor crystal are held together by *covalent bonding*. Although silicon and germanium have only *four valence electrons* per atom, the covalent bonding allows each one to 'share' four more with four neighbouring atoms (Fig. 11.5).

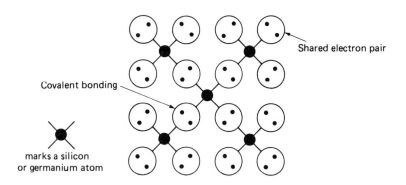

Figure 11.5

The lattice thus formed is *stable* and has *no free electrons*, i.e. the material is behaving like an insulator. However, under normal temperatures atomic vibrations cause some of the covalent bonds to be *broken*. Also, defects—*dislocations*—in the lattice and *impurities* provide free electrons. With increasing temperature, these properties allow the material to become a *conductor*, albeit a *weak conductor*.

The liberation of an electron from a covalent bond leaves behind a positive vacancy—*the hole*. On application of an electric field, electrons move from hole to hole in the direction of the higher potential, giving the effect of current and the 'movement' of holes.

Donation and acceptance

Addition of an impurity can make *more holes or electrons* available and so affect the properties of the semiconductor. These effects are considerable even with impurities of one in a million, since only one in ten-thousand million semiconductor atoms supplies a conduction electron.

If, for example, an impurity atom having *five valence electrons* replaces a silicon atom in the lattice, it has one valence electron left after bonding to neighbouring silicon atoms. The electron then acts as a *free electron* and the impurity is termed a **donor** (Fig. 11.6). The same occurs in a germanium lattice. Examples of donors include antimony, arsenic and phosphorus.

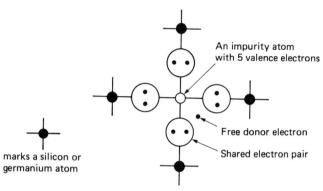

Figure 11.6

The semiconductor is said to be of **n-type** because conduction is mainly by negative charge movement. Donation can be represented by the formation of extra donor energy levels in the forbidden energy gap (Fig. 11.7).

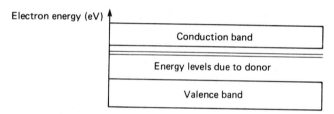

Figure 11.7

The difference between these extra levels and the conduction band is about 0.1 eV.

A similar set of energy levels is formed just above the valence band when **acceptor impurities** are added. These have *three valence electrons* per atom, e.g. gallium, boron and aluminium, and take part in covalent bonding with *only three* silicon or germanium *atoms*. In the region of the fourth bond is *one valence electron* and *a hole*. Such a hole can be filled by electrons from neighbouring silicon or germanium atoms (Figs 11.8 and 11.9).

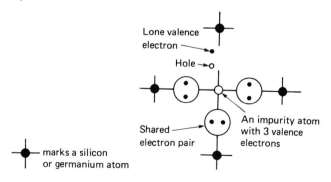

Figure 11.8

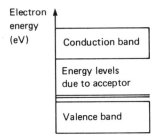

Figure 11.9

The result is an *increase* in the number of *holes* which then become responsible for most of the conduction. Such a semiconductor is termed **p-type**.

Carriers

In the n-type and p-type semiconductors, the electrons and holes respectively are called the **majority carriers**. The holes and electrons respectively are called the **minority carriers**. Carriers carry charge.

Addition of impurities, called **doping**, increases the number of majority carriers and also in effect decreases the number of minority carriers by making recombination more likely. Semiconductors that are doped are said to be **extrinsic**.

p–n junction diode

Devices consisting of two types of semiconductor, or a semiconductor and a metal in contact, which allow conduction mainly in *one direction* are called **semiconductor diodes**.

For example n-type and p-type semiconductors are formed side by side in the same crystal of silicon or germanium (Fig. 11.10).

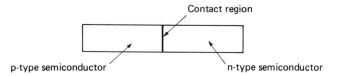

Figure 11.10

Initially both sides of the contact region are *neutral*. However, some of the free or conduction electrons in the n-type side move or *diffuse* into the p-type side. Here they begin to fill up the holes near the contact region leaving an overall negative charge there. At the same time some holes diffuse into the n-type side where they are filled by electrons, and so an overall positive charge is set up there.

As a result an *electric field* is formed across the boundary of the two types of semiconductor. This field gets stronger and opposes the diffusion of further majority carriers until the diffusion is stopped altogether. The boundary area then acts as a **potential barrier** and is called the **depletion layer** (Fig. 11.11). It has a high resistance because there are very few carriers there.

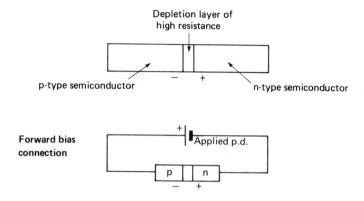

Figure 11.11

When a p.d. is applied across the diode, which makes the p-type side less negative, or even positive, with respect to the n-type side, a *current* is created. However, if the applied p.d. is reversed, the electric field preventing carrier movement at the depletion layer gets stronger. (A *small current* does exist from the movement of *minority carriers*.)

In the former case the diode is said to have **forward bias** and in the latter case it has **reverse bias**. The forward bias connection is shown in Fig. 11.11 and turning the battery around would give reverse bias.

Diode characteristic curves

The *characteristic* curves (Figs 11.12 and 11.13) show current versus applied p.d. for a germanium diode in both forward and reverse bias. The two graphs are plotted using *different scales*.

Forward bias

In the forward bias (Fig. 11.12), current becomes appreciable when the applied p.d. is greater than the p.d. across the depletion layer.

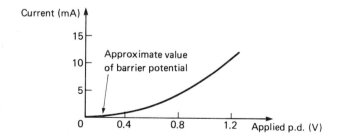

Figure 11.12

Reverse bias

In the reverse bias (Fig. 11.13), an appreciable current is obtained when the electric field in the depletion layer is strong enough to promote

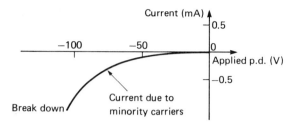

Figure 11.13

electrons into the conduction band, although this is damaging to the diode (break down).

The diode itself can be made up from a wafer of a certain type of semiconductor soldered onto a metal base. Into the wafer is projected a pellet of impurity which converts part of it into the opposite type of semiconductor. Both pellet and base are then attached to wires.

Rectification

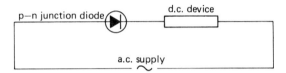

Figure 11.14

The action of the diode can be usefully applied in the *rectification* of a.c. for a d.c. device as in the circuit in Fig. 11.14. The current produced in the device is varying, but in one direction (Fig. 11.15).

Figure 11.15

A similar graph can be drawn for the p.d. across the device.

A steadier current can be obtained by introducing a *storage capacitor* across the device. This is called **capacitor smoothing**. The current through the diode charges up the storage capacitor until the p.d. across the capacitor is equal but opposite in direction to the peak a.c. p.d. Current in the diode then stops. The capacitor discharges through the d.c. device, and the p.d. across it drops slightly as it begins to do so. This enables current once again to pass through the diode, which again completely charges up the capacitor, and the process repeats.

Example 11.1

The circuit in Fig. 11.16 consists of a step-up transformer connected to a full wave rectifier. The transformer primary has one-quarter of the turns which the secondary has, and is applied across an a.c. source of 50 V. If no energy loss occurs either in the transformer or in the rectifier circuit diodes, then (i) explain how the rectifier works, (ii) find the rms value of the current passing through the resistors, and (iii) find the peak value of the p.d. across the 35 Ω resistor and also across the other resistor.

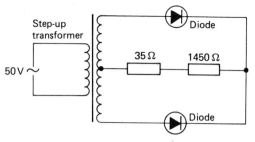

Figure 11.16

(i) The centre tap on the secondary essentially cuts the p.d. produced across the secondary into two equal parts. At any instant one part will allow current to pass in the correct direction through one of the diodes and through the resistors. However, the other diode will not allow this direction of current. In the next cycle of p.d. the situation is reversed and the other diode will not permit current to pass while its partner will, and so on. Thus, current passes in one direction only through the resistors, although there is a variation in its size. (The current is not being smoothed.)

(ii) Applying the transformer equation, then

$$\frac{V_s}{V_p} = \frac{N_s}{N_p} = 4$$

Therefore

$$V_s = 4V_p = 4 \times 50 = 200 \text{ V}$$

Thus, the peak p.d. across the resistors is 100 V and so the peak current passing in the resistors is given by

$$I = \frac{V}{R_1 + R_2} = \frac{100}{35 + 1450} = 0.067 \text{ A}$$

Therefore

$$\text{rms value of current} = \frac{0.067}{\sqrt{2}} = \underline{0.047 \text{ A}}$$

(iii) The peak value of the p.d. across R_1 is

$$\frac{100R_1}{R_1 + R_2} = \frac{100 \times 35}{35 + 1450} = \underline{2.36 \text{ V}}$$

Thus,

$$\text{peak p.d. across the other resistor} = 100 - 2.36 = \underline{97.64 \text{ V}}$$

Capacitor smoothing circuit

See Fig. 11.17. Similar graphs can be drawn for p.d. versus time.

Bipolar transistor

A transistor consists of three layers of different types of semiconductor formed in a single crystal. One of the methods in making a transistor involves the immersion and slow withdrawal of a germanium seed crystal from molten germanium. As the crystal is raised, different impurities are

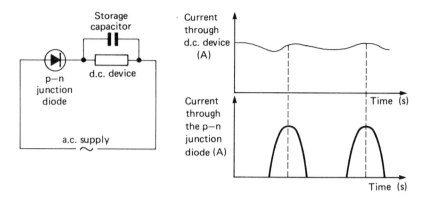

Figure 11.17

added to the molten germanium, enabling different types to be formed in the 'growing' crystal.

An npn transistor is *rather like* two p–n junction diodes placed together 'back-to-back', where one is forward biased and the other reverse biased. There are three parts: the **emitter**, **base** and **collector** (Figs 11.18 and 11.19).

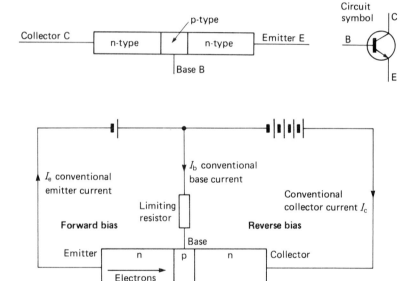

Figure 11.18

Figure 11.19

The region between the emitter and base is **forward biased** and electrons flow from the former to the latter. It has a *low resistance*. The base–collector region is **reverse biased** and will not permit electrons to go from the collector to the base. It therefore has a high resistance. However, it does permit electrons to go from the base to the collector.

Most of the electrons from the emitter reach the collector since the base is thin, but some recombination of electrons and holes in the base does indeed take place. As a result a small base current I_b is set up and its relation to the other currents (I_c and I_e for the collector and emitter respectively) is

$$I_e = I_b + I_c$$

I_e is about 5 mA for an emitter–base p.d. of 0.6 V, while I_b is about 50 μA.

Therefore

$$I_e = I_c \quad \text{(approx.)}$$

The movement of electrons between regions of low and high resistance give the **transistor** its name: **trans**fer of re**sistance**. There is a *strong* relationship between the collector current and the base current and small variations in the latter produce much greater variations in the former.

Thus, the addition of weak a.c. to the base current produces greatly *amplified* variations on top of the steady collector current.

Characteristics of a transistor in the common emitter mode

Graphs showing current and p.d. relationships for a transistor are called its **characteristics**. In the common emitter setup, called a *mode*, both collector and base p.d.'s are taken with respect to the *emitter*.

Circuit for common emitter mode

See Fig. 11.20.

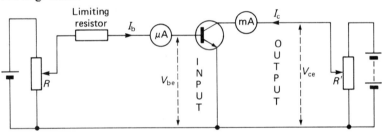

Figure 11.20

A **limiting resistor** is used to ensure that the base current I_b never exceeds the maximum value which the transistor can take.

The p.d. across the base and emitter V_{be} and that across the collector and emitter V_{ce} are measured with an oscilloscope to prevent loss of current.

Input (or base) characteristic

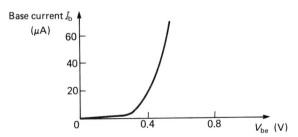

Figure 11.21

See Fig. 11.21. When V_{ce} is kept *constant* and I_b is varied by adjusting the variable resistor R, an *input characteristic* is obtained for measurements of V_{be}. The value of I_b increases rapidly when V_{be} exceeds 0.6 V and the curve resembles that for a p–n junction diode with *forward bias*. Variation of V_{ce} causes the curve to shift only slightly sideways.

Output (or collector) characteristic

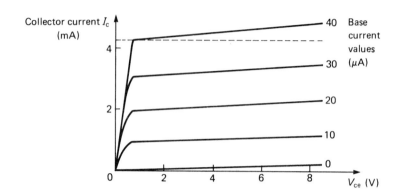

Figure 11.22

When V_{be} is kept *constant* (giving a particular value of I_b, e.g. 10 μA, 20 μA or 30 μA), the variation of I_c and V_{ce} is as shown in Fig. 11.22. Care has to be taken for small readings of V_{ce}, as I_c changes fairly rapidly, and the value of I_b does not alter because of change in V_{ce}.

Beyond the range 0.0 V to 0.2 V for V_{ce}, I_c is determined mainly by I_b and *not* by V_{ce}. Hence the flat parts of the curve, called regions of **current saturation**. These occur because I_c is due to charge carriers from the emitter, i.e. V_{ce} has little effect on I_c since it is not V_{ce} that produces the charge carriers.

When the input current I_b is *zero*, a small value of I_c is observed. This is the **leakage current** and is due to the movement of holes from collector to the base. The *reverse bias* at that p–n junction makes this possible, although the current is small since these are minority carriers. See Fig. 11.23.

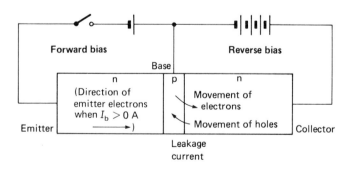

Figure 11.23

This movement of the holes is equivalent to a movement of electrons in the opposite direction.

The slope of the flat part of the curves gives the *output resistance* of the transistor which is great because of the *reverse bias*.

Transfer characteristic

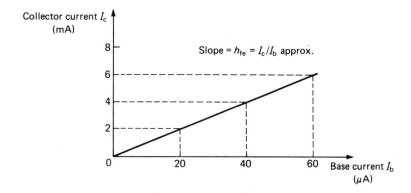

Figure 11.24

When the value of V_{ce} is kept *constant*, the graph in Fig. 11.24 illustrates the variation of I_c with I_b. It can be obtained from the output characteristics and hence the dashed lines on the graph. The plot is really a slight curve, although a straight line is usually drawn.

Hybrid parameters

In the analysis of complicated circuits, it is usual to set up a simplified or equivalent theoretical circuit. The simple theoretical circuits for the action of a transistor are described in terms of certain values. These are the *h* or **hybrid parameters**. Two of them have the symbols h_{fe} and h_{ie}.

h_{fe} is called the **static value** of the forward **current transfer ratio** or the **static current gain** for the common emitter mode. It is obtained from the gradient of the *transfer* characteristic. The term static applies to d.c. operation, as opposed to dynamic, which applies to a.c. operation.

$$h_{fe} = I_c/I_b \quad \text{(approx.)} \quad \text{(no units)}$$

h_{ie} is the **input hybrid parameter** for the common emitter mode. It can be obtained from the input characteristic as the reciprocal of the gradient.

$$h_{ie} = \frac{\delta V_{be}}{\delta I_b}$$

By selecting a *particular point* on the input characteristic and finding the ratio V_{be}/I_b, a *static* value of input resistance is obtained. However, since the input is an alternating value superimposed upon a steady d.c. value when the transistor is used in amplification, h_{ie} is more useful.

Characteristics of a transistor in the common base mode

A similar set of characteristics for the transistor can be found for the common base mode, where the p.d.'s are measured with respect to the *base*.

Circuit for common base mode

See Fig. 11.25.

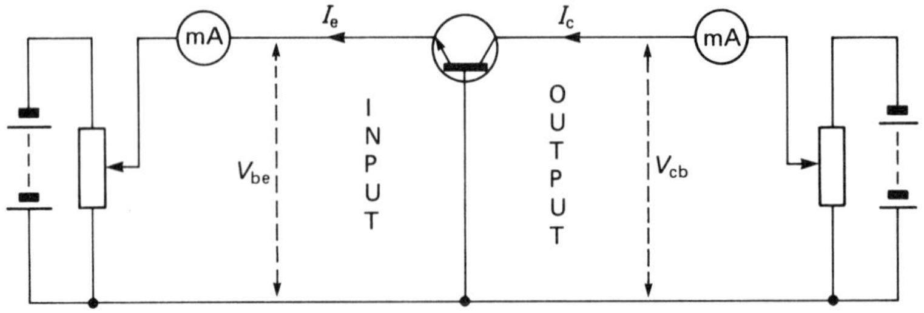

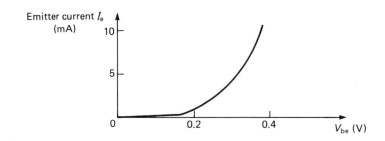

Figure 11.25

Input characteristic

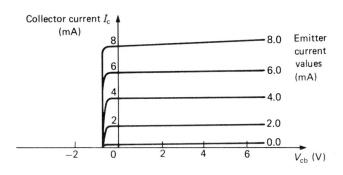

Figure 11.26

The characteristic (Fig. 11.26) is again like that of the *forward biased* diode as this is basically what the base–emitter junction is. The input resistance is given by $\delta V_{be}/\delta I_e$.

Output characteristic

Figure 11.27

Again flat curves are produced (Fig. 11.27). The curve for $I_e = 0\,A$ is that given by a *reverse biased* diode, and on the whole all the curves are *flatter* than for the common emitter mode. Even when V_{cb} is zero, the collector can gather electrons from the base because of the depletion layer p.d. See Fig. 11.28.

The depletion layer p.d. is small and only a small reversed V_{cb} is needed to make I_c zero.

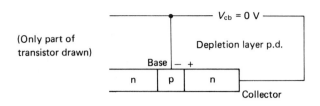

Figure 11.28

Transfer characteristic

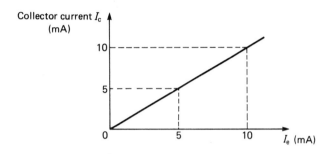

Figure 11.29

Since

$$I_e = I_c + I_b$$

and I_b is comparatively small, then $I_e = I_c$, approximately, and so the transfer characteristic for the common base mode is a straight line graph, almost passing through the origin, and with a gradient of nearly one (Fig. 11.29).

Some typical values

Quantity	Typical values
Static current gain	50 to 100
Input resistance	
common emitter	1 kΩ to 3 kΩ
common base	50 Ω to 100 Ω

Amplification

Amplification is usefully employed in many devices, e.g. hi-fi equipment, electrical measuring devices, and electric detectors. There are a number of types of amplifier but one of the simplest is the **common emitter** class A a.c. amplifier (Fig. 11.30).

For zero input the base or bias current I_b is given by V/R, approximately, where R is the bias resistance and V is the p.d. applied by cells. This is an approximation because part of V is applied across the base–emitter (V_{be}), and theoretically this is about 0.6 V. The current I_b enables the transistor to be properly *biased*, i.e. the emitter junction with forward bias and the collector junction with reverse bias.

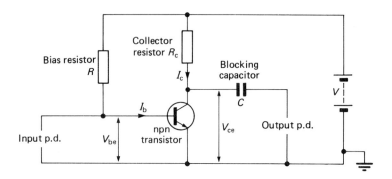

Figure 11.30

The steady p.d. across the collector resistor:

$$V_c = I_c R_c = h_{fe} I_b R_c$$

Also,

$$V_c = V - V_{ce} = h_{fe} I_b R_c$$

Therefore

$$V_{ce} = V - h_{fe} I_b R_c$$

When a weak a.c. signal is added to the bias current I_b, large variations in I_c occur. This means that V_c *varies greatly* and in turn so does V_{ce}, the output p.d.

Finally, a *blocking* capacitor C is used to remove the steady p.d. from V_{ce} which is there with or without the a.c. input. The gain of the amplifier is defined as the ratio of the change in output p.d., δV_o, to the small change in input p.d., δV_i, i.e.

$$\text{gain} = \frac{\delta V_o}{\delta V_i}$$

Now,

$$h_{ie} = \frac{V_{be}}{I_b} = \frac{V_i}{I_b}$$

Therefore

$$\delta V_i = h_{ie}\, \delta I_b$$

Also,

$$V_{ce} = V_o = V - h_{fe} I_b R_c$$

Thus,

$$\delta V_o = -h_{fe} R_c\, \delta I_b$$

(as V is constant). Then

$$\text{gain} = \frac{-h_{fe} R_c\, \delta I_b}{h_{ie}\, \delta I_b}$$

So

$$\boxed{\text{gain} = \frac{-h_{fe} R_c}{h_{ie}}}$$

The minus sign indicates that the output and input are in antiphase.

This type of gain is called **voltage gain**.
The **current gain** is given by

$$\text{current gain} = \frac{\delta I_c}{\delta I_b} = h_{fe}$$

Switching

The fact that small changes in the base current of a transistor cause large changes in the collector current is used in electronic switching.

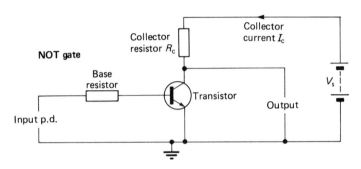

Figure 11.31

A particular type of switch whose action is similar to that of a relay switch is shown in Fig. 11.31.

Example 11.2

The p.d. across the base and the emitter of a transistor connected in the common emitter mode rises from 1.2 V to 1.5 V. If at the same time the base current increases from 120 μA to a value of 175 μA, determine (i) the value of the static input resistance and (ii) the value of the input hybrid parameter h_{ie}.

(i)   The static input resistance is given by

$$R = \frac{V_{be}}{I_b} = \frac{1.2}{120 \times 10^{-6}} = \underline{1 \times 10^4 \ \Omega}$$

(ii)   The input hybrid parameter for the common emitter mode is given by

$$h_{ie} = \frac{\delta V_{be}}{\delta I_b} = \frac{1.5 - 1.2}{(175 - 120) \times 10^{-6}} = \underline{5.5 \times 10^3 \ \Omega}$$

Example 11.3

The p.d. across the collector and the base of a transistor connected in the common base mode rises from a value of 7.5 V to 11.5 V. If at the same time the collector current increases by an amount of 18 μA, calculate the output resistance assuming no change in emitter current occurs.

The output resistance for the common base mode will be given by

$$R = \frac{\delta V_{cb}}{\delta I_c} = \frac{11.5 - 7.5}{18 \times 10^{-6}} = \underline{2.2 \times 10^5 \ \Omega}$$

Example 11.4

The constant p.d. supplied by the cells in a common emitter amplifier with respect to the earth rail is $+5$ V, and the base resistor has a value of 75 kΩ. Find the value of the collector load resistor which will enable saturation to occur for a minimum value of the base–emitter p.d. of 2.5 V. (The static current gain of the transistor is 75.)

The static current gain is given by

$$h_{fe} = \frac{I_c}{I_b}$$

Now,

$$V_{ce} + R_c I_c = 5 \text{ V}$$

and for saturation

$$V_{ce} = 0 \text{ V}$$

Thus,

$$R_c I_c = 5 \text{ V}$$

and so

$$R_c I_b h_{fe} = 5 \text{ V}$$

For saturation to occur, the base–emitter junction has to be forward biased and so the input resistance is small. Therefore

$$I_b = \frac{V_{be}}{R_b} \quad \text{(approx.)}$$

Therefore

$$\frac{R_c V_{be} h_{fe}}{R_b} = 5 \text{ V} \quad \text{and} \quad R_c = \frac{5R_b}{V_{be} h_{fe}}$$

Hence

$$R_c = \frac{5 \times 75 \times 10^3}{2.5 \times 75} = \underline{2 \times 10^3 \ \Omega}$$

Saturation is the condition of the transistor when no matter how high the base current becomes, the collector current is unchanged.

Switch theory

When the transistor base is nearly at zero potential with respect to the emitter, the output p.d. is high—approximately V_s (see Fig. 11.31). This situation is called **cut-off** and I_c is zero. This type of switch is called an **inverter** (or **NOT gate**), because the output is high for a low input. When the input is high, the output is well below 0.3 V and the situation is called **saturation** since, no matter how high the base current becomes, the value of I_c is unaltered. These two different conditions are explained from the equation

$$\boxed{V_s = V_{ce} + R_c I_c}$$

When input is small, I_b is small and consequently I_c is small, and therefore $V_{ce} = V_s$, approximately. When input is large, $R_c I_c$ becomes large, and since V_s is constant, the output V_{ce} falls to near zero.

The cut-off and saturation regions are shown on an output p.d. versus input p.d. graph (Fig. 11.32).

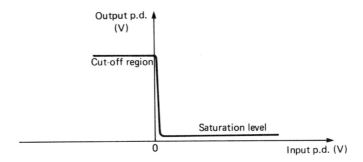

Figure 11.32

Logic gates

Voltage signals which have only two possible values, e.g. zero and another fixed value, can be used to represent numerical digits. The actual values of the signals are unimportant, and digital circuits which employ them need only have *two stable* conditions. These are represented by a transistor carrying current or not. They are termed state 0 and state 1, or off and on, respectively.

As just two digits are available, the *binary* number *system* is used in these circuits. Circuits which perform operations on digital signals and have two or more inputs and one output are called **logic gates**.

Truth tables

A list of all the possible combinations of input values showing the true output value for each combination for a logic gate is called a **truth table**. It describes the logic function of that particular gate.

NOT gate

Truth table for NOT gate.

Input	Output
1	0
0	1

Application of the NOT gate

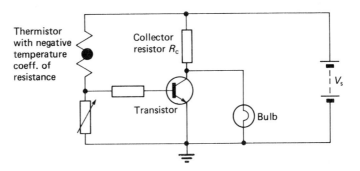

Figure 11.33

When there is a marked fall in temperature, the bulb in the circuit in Fig. 11.33 comes on. This is because the resistance of the *thermistor* increases and the input p.d. drops to a low value. The NOT gate is now in the condition defined by the last part of its truth table.

It is usual to combine the NOT gate function with other logic gate functions. This is because the NOT gate is basically a common emitter amplifier, and *amplification* of signals in it preserves the amplitude of the signals as they pass through a network of multiple logic gates.

NOR gate

A NOR gate is a NOT gate with *two or more* resistive inputs (Fig. 11.34). In this case only if all the inputs are low then the output is high. (The output is in state 1 if neither one *nor* the other inputs is also in state 1.)

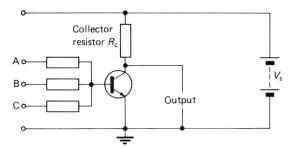

Figure 11.34

Truth table for NOR gate

Inputs A	B	C	Output
0	0	0	1
1	0	0	0
0	1	0	0
0	0	1	0
1	1	0	0
0	1	1	0
1	0	1	0
1	1	1	0

OR gate

The OR gate is the complete opposite to the NOR gate. That is, if either one *or* the other inputs is high then the output will be high.

This would mean for the above truth table, for an example of a NOR gate, that all the outputs in state 0 would have to become outputs in state 1 and that the output in state 1 would have to become an output in state 0. Such a process is called **inversion** and is achieved by following the NOR gate with a NOT gate.

An example of an OR gate with *two* inputs is given (Fig. 11.35), as well as its truth table.

OR gate circuit

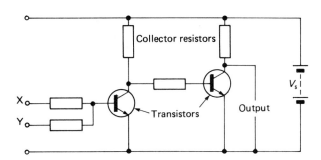

Figure 11.35

Truth table for OR gate

Inputs		
X	Y	Output
0	0	0
1	0	1
0	1	1
1	1	1

AND gate

In this type of gate the output is only high when one *and* the other inputs are high. It can be constructed by placing a NOT gate at each of the resistive inputs to a NOR gate (Fig. 11.36).

The NOR gate will only provide a high output if all of its inputs are low. These are indeed low, only when the inputs of the NOT gates, which are attached to them, are high.

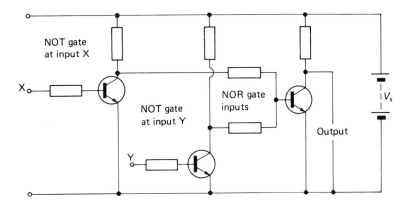

Figure 11.36

Another name for the AND gate is the **coincidence circuit**. It is useful in nuclear physics where two or more events which arise simultaneously need to be recorded.

NAND gate

This is the complete reverse of an AND gate and can be made by placing a NOT gate after an AND gate.

Truth table for AND gate

Inputs X Y	Output
0 0	0
1 0	0
0 1	0
1 1	1

Truth table for NAND gate

Inputs X Y	Output
0 0	1
1 0	1
0 1	1
1 1	0

Negative logic

The use of signals which are at a small positive potential to represent state 1 and signals which are at zero potential to represent state 0 is called **positive** logic. If these representations are reversed, then **negative** logic is obtained, as state 1 becomes more negative than state 0.

Thus, for example, the truth table for an AND gate would have to be rewritten as follows.

Truth table

Inputs X	Y	Output
0 (high)	0 (high)	0 (high)
1 (low)	0 (high)	1 (low)
0 (high)	1 (low)	1 (low)
1 (low)	1 (low)	1 (low)

The truth table for the *AND gate* with *negative* logic has become the truth table for the *OR gate* with *positive* logic. Thus, a NOT or **inverter** gate can be considered as a logic exchanger. For instance, if a NOT gate is placed in front of an OR gate, the signals reaching the OR gate are inverted and a negative logic AND gate is obtained. The addition of another inverter gate onto the end of the combination produces a positive logic AND gate.

Logic circuits

The transmission of the several bits of a complete binary number can be accomplished in two ways.

Firstly, by transmission from a single source in the form of equally spaced pulses of constant amplitude. This is done sequentially starting with 2^0. Each digit—either a 0 or 1—is called a **bit**, which is short for **binary digit**. It is called *serial* representation (Fig. 11.37).

Secondly, pulses representing increasing powers of two are created simultaneously, and this is termed *parallel* representation.

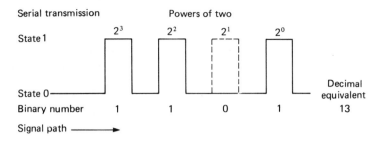

Figure 11.37

Serial transmission has the advantage that it requires only one signal path, whereas the advantage of parallel transmission is in the time needed for transfer of the number—only one bit's worth.

Different types of logic gates are used in electronic circuitry. DL (diode logic) gates employ only diodes but have the disadvantage of signal weakening in complex networks. RTL (resistor–transistor logic) or TTL (transistor–transistor logic) gates are more commonly used.

Circuit symbols for gates

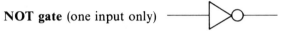

NOT gate (one input only)

The symbols in Fig. 11.38 represent gates with three inputs.

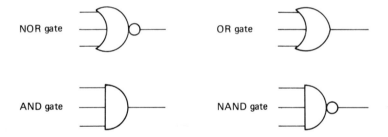

Figure 11.38

Example 11.5

Lifts are used on an aircraft carrier to raise aircraft to deck level, for launch. The lift is operated by switching the lift button on, and if the aircraft is in the correct position it will be raised up to the deck. If the aircraft is incorrectly positioned on the lift, the lift motor will not activate. On reaching the deck level, an electrical contact is made which

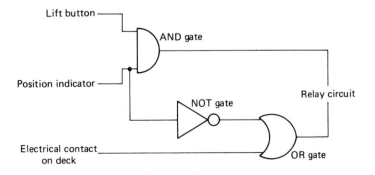

Figure 11.39

disengages the lift motor. Design a system of logic gates which will control a lift mechanism in the above described manner.

See Fig. 11.39. The conditions which initiate the raising of the aircraft, i.e. the depression of the lift button and the correct positioning of the plane, determine the action of an AND gate, which produces an output when the conditions arise simultaneously. The conditions which prevent the plane from being raised or being raised further affect the operation of an OR gate. Since the positioning of the plane determines also the effect of the AND gate, a NOT gate is interposed between that input and the OR gate. Both OR and AND gates then feed a relay switch circuit which will allow current to be switched to the motor, thus activating it. The relay will operate only if there is current from the AND gate, and not if there is current from the OR gate.

Index